Secrets of the skeleton

Form in metamorphosis

L.F.C. Mees

Secrets of the skeleton

Form in metamorphosis

SteinerBooks
Anthropsophic Press

This book first appeared in 1980 as *Geheimen van het skelet: Vorm en metamorfose* published by Uitgeverij Vrij Geestesleven of Zeist, Holland. It was translated by the author and edited by Ellen Bohr and David Adams.

Library of Congress Cataloging in Publication Data

Mees, L.F.C.
Secrets of the skeleton.

Translation of: Geheimen van het skelet.
1. Human skeleton. 2. Human evolution.
3. Fossil man. I. Title.
QM101.M3613 1984 611 84-12360
ISBN 0-88010-087-7 (pbk.)

Printed in the United States of America

Contents

Chapter 4

Viewing our discovered phenomena in a new light

Chapter 5

The tripartite human being

Chapter 6

Metamorphosis and evolution

Preface

As a youngster, I often went roaming through the sand dunes in North Holland with groups of boys who spent summer holidays camping there. Many times we found rabbit skulls and sometimes parts of other small skeletons. We all took great delight in them, admired them, and took them home. I think that quite a few people in Holland still have a little rabbit skull on display somewhere in their house.

Many people, however, feel that skeletons have a rather sinister connotation—especially the human skeleton. Up to a point, that is understandable. Death is often depicted as a skeleton with a scythe. Death, skeleton, bones—this is an obvious association. This is also the reason why many people do not get to know the typical characteristics of various bones. Faced with a heap of large and small bones, most of us will see only that. How much all of this changes when one begins to *distinguish* them.

For this purpose one should not only *look* at the bones of a human skeleton, but pick them up, touch them, stroke the flat, round and angular surfaces, in order to experience the intrinsic modeled beauty of the many varied shapes. We must learn to see not only with our eyes, but also with our fingers. This in turn enhances our ability to notice things that at first glance escape us. We begin to discover things that reveal themselves in the true sense of the word only after we have learned to distinguish the essential from the accidental.

In this book we are not dealing, as is usually the case, with statements that are, in my opinion, established facts. Here we have to do with such questions as: Is it possible to see this in such a way? In a discussion in which the artistic and the scientific elements are so closely intertwined, it cannot be otherwise.

I want to stress, however, that the comparisons and conclusions that follow should not simply be called subjective. The artistic part that lives in some form in each human being can be used as a means of observation. Thus, it is possible to observe things that would escape those who take the purely analytical, scientific approach. To view a landscape or a painting one must stand at a distance. One comes too close by using a magnifying glass. Many more details are seen, but the sight of the whole is lost. When one goes to the moon one loses sight of the moon; that is to say, one loses the relation to the surroundings that gives certain details their true value. With it, one loses a certain source of understanding.

The great Professor of Anatomy and Anthropology, Louis Bolk, who first taught me osteology, said in his book *Brain and Culture*:

> We are in the habit of investigating life through magnifying glasses, of thus bringing otherwise invisible matter within our field of vision; how different, how much larger our concept of life would be if it were possible to study it through minimizing glasses and thus bring within our field of vision matters that are beyond the reach of the eye, taking as a goal of our studies the cohesion of the phenomena rather than the analysis of matter, as we are doing now.

I have tried throughout the book to apply this method to the study of the metamorphoses of the skeleton.

Finally, it will be clear that the experiences I am going to describe are but a selection taken from a much wider field still open to exploration. It is hoped that readers attracted to the examples will carry on and open up new land in this hitherto unexplored area.

It is of prime importance to discover a certain order, a plan, in the multitude of shapes. To achieve this we must study the skeleton as a whole. For our purpose it is also necessary to study the shape of a number of bones in a new way and by mutual comparison.

I began this study in the thirties. Some artists of my acquaintance offered their help by making drawings. There is a difference between a drawing and a photograph. A draftsman stresses points he considers of particular interest. This subjective approach carries with it the danger of suggestion. A photograph can never show things that are not there. Therefore I have, wherever possible, placed the drawings of two artists, each with his own particular technique, next to a photograph, for the sake of comparison. After each photograph follows first a drawing by Ing. W. G. ten Houten de Lange (former director of the Zeiss planetarium in The Hague) and then a drawing by Rupert van der Linden. The ancient drawings are made by Wandelaar in a textbook of anatomy by Bernard Siegfried Albinus (1697-1770).

Driebergen, August 1979 *L.F.C. Mees*

Pl. 1. Natural stone.

Pl. 2. Stone axe head.

Pl. 3. Stone relief.

Pl. 4. Tooth of a sperm whale.

Chapter 1
Diversity of shapes

Before studying the actual shapes of the skeleton, it will be useful to consider shapes in general, including some from the non-living world around us.

Plates 1-4 show a natural stone, a stone axe (utensil), a small stone relief, and a bone (the tooth of a sperm whale). All four objects are said to have a shape. In all four cases, shape has been achieved in a different way. The first stone has been shaped by nature. Although certain laws have played a part, the result can nevertheless be called haphazard. Therefore, we are dealing here with a shape that is mainly a contour. This of course is also true of the other three; but in their case it is a question of a definite, though different, point of departure.

The axe is pragmatic; its shape can be called a construction. All utensils as such are constructions. In a work of art we find an idea, an attempt at beauty. Here one can use the word composition. Finally, in the case of the bone, we use the word creation, which applies to the outside as well as to the inside. Thus we have a progression: contour, construction, composition, creation.

In the production of the axe and the art object we can speak of viewpoints, but does the same apply to shapes created by living nature? A viewpoint in this case means that which coordinates all the component parts of a shape. The viewpoint is well known when we are the creator of the shape.

But what about shapes like those of bones? We recognize without difficulty their unity, we are aware of the natural laws that give them their cohesion, and we feel compelled to consider these shapes the result of a creation. We know the creator of the axe and of the work of art is the human being. The question arises: How should we conceive of the creator of bones?

I think there is a very good reason for this question, one that is all too often overlooked. We call (or used to call) all natural shapes creations. Why?

What inspired the human being to do this? The fact that *he himself* is a creative being, the only creative being on earth. It is right to point out that the animals do not *create*, in spite of their great instinctive gifts. Animals *repeat*.

Being a creator, the human being recognizes a natural shape as a creation and is capable of considering the relationship between creator and creation in nature as an object of study. This is justified because he works from within his own experience. Here one could ask: Should we conceive of God, or Creator, as anthropomorphic (human)? The answer is, on the contrary, that the creative human being is to be considered as theomorphic (godly).

We can compare the latter relationship between the creative forces and matter with the equivalent relationship between a work of art and a technical form. The artist and the technician both work from the outside on the material. The artist will be inclined to seek a closer relationship with the material than will the technician, who is primarily concerned with its utility from a technical point of view. Thus the sequence, axe—work of art—bone, becomes even more a true progression. There is an ever stronger tie between the one who works, who creates, and the material.

When we study the shape of a part of the skeleton, we must remember that the bone was once permeated with the creative shaping principle, which withdrew after death. We will again refer to these shaping forces in Chapter IV.

We must be well aware that we are going to study a world which we shall approach, so to speak, from within its script. At the same time, we must take care to leave out attributes that can be ascribed to human creations, such as utility. We are going to study the shapes of the human skeleton in such a way that we will try to find the inner correlation of many, apparently quite different, shapes. We will try to find the hidden theme, just as we would do in a piece of music.

The method we refer to here has been applied by Goethe in his extensive research on plants and animals. The results are described in his *Metamorphosis of Plants* and *Metamorphosis of Animals*. Although the latter is not quite satisfactory, the former is a masterpiece!

First, it will be necessary to examine more extensively the concept of metamorphosis (literally, change of shape).

Chapter 2
Metamorphosis in the vegetable, animal and human realms

Metamorphosis in the plant

The phenomenon of metamorphosis, as described by Goethe, implies that a creative principle calls forth various shapes in a certain way. This certain way consists of a tendency toward development in these shapes. Development is achieved by reaching a final point that is in complete contrast to the point of departure. What is meant by this can best be illustrated by close observation of the metamorphosis of the plant.

Stem, leaves and flowers are the characteristic parts a higher plant shows above ground. Even before Goethe, naturalists were struck by the fact that a leaf sometimes shows partial characteristics of a petal. The reverse is also seen in petals that have a green part. Stamens can become partial petals again, fruit leaves can grow into other leaflike shapes, and so on. It became clear to Goethe that all these plant organs—stem, calyx and leaves, stamens, and carpels—must have the same underlying formative principle. Moreover, he noticed a steady increase in quality, specialization, delicacy, and beauty from the first leaves to the appearance of the flower. One can use other words and speak of "hidden" and "revealed." Such expressions are justified, especially if one thinks of the contrast between seed and flower. The plant reaches its final point in the flower, where it can reveal itself no further.

Goethe called this process *Steigerung* ("intensification"); seed and flower were to him a polarity. Polarity and intensification are the laws most fundamentally connected with the concept of metamorphosis.

But to ascertain the existence of relationships between shapes in various plant organs is not enough to reveal the essence of metamorphosis. When observing a fully grown plant—a rose, for example—we may observe that in the course of time stem, leaves, and petals have been put forth. This unfolding growth finally results in a flower. Then further growth stops. The plant repeatedly reaches its final stage in the flower.

In reality the development of a plant runs a different course. The growth of the stem of a plant always begins with the formation of leaves. Even the first germination begins with the development of the seed lobes, or seed leaves, which are essentially already present. As soon as an inclination to develop is felt in this region of the leaf bud (through the warmth and light of spring), the stalk beneath the bud begins to grow and becomes a stem.

Meanwhile, new buds have formed in the leaf axils. As soon as spring's influence stimulates these new buds to develop further, the growth of a new stem follows. This stem is, therefore, the original stalk of the bud. In this way a plant "edges its way up" from leaf node to leaf node. With each step more or less new leaf shapes also appear. The same principle is manifested again and again in a slightly altered guise. Thus the plant can be seen as a greater or smaller number of leaf-bud and stem articulations—ending in a flower-bud and stem—that are "roused" by seasonal influences. The plant is a seasonal being.

This is likewise related to the polarity of seed and flower. Its image is the contrast between earth and heaven (cosmos). It is the latter that produces the intensification. As has already been said, at each successive step from bud-stem unit to bud-stem unit the leaves will look slightly different, showing the phenomenon of intensification. That this process may vary from plant to plant does not contradict the principle. From this point of view, one can say that the environment constantly stimulates new shapes in the plant.

It is often asked if variations are also metamorphoses. The answer is that they are not, in the sense of our definition. When can we speak of variations? Let us imagine a large field, filled with flowering daisies. Each plant shows in its growth the metamorphosis we have described. When one compares the flowers, petals or leaves of several plants at the same level, one notices that no two daisies are exactly alike. These are variations. Why not metamorphosis? Because here neither polarity nor intensification appears.

In the same way a number of specimens of one species of butterfly can show variations. By thinking of the field of daisies where we observed and compared variations and metamorphoses, we conclude that variations show a "next to each other" while metamorphoses show an "after each other."

For the sake of completeness we must speak briefly about the shapes we encounter in technology and art. Especially in technology we find many variations of the same invention. If, for example, all cars looked alike, interest in them would soon wane. Producing many makes of cars maintains interest. If one places next to each other the many cars produced since their invention nearly a century ago, we clearly see a metamorphosis. The same idea has undergone such a development that one can talk of intensification in the sense of *Steigerung*. There are great contrasts, though we have not yet seen the end.

There are artists who have worked on the same theme all their lives. In these cases an idea presents itself under a different guise because the artist is a developing being, and this finds expression in his work. A splendid example is Monet, who repeatedly used the theme of water lilies in his paintings.

Metamorphosis of the animal

In making the transition from metamorphosis in the plant world to that of the animal world, many of us immediately think of the well known examples of the butterfly and the frog. The Dutch language has a unique expression for the metamorphosis that is studied in these animal groups: *gedaanteverwisseling* "[ex]changing of form." Everyone knows the cycle in the insect world: egg, caterpillar, chrysalis, butterfly, as well as that of the amphibians: egg, tadpole, frog.

The cycle egg, caterpillar, chrysalis, butterfly bears a charming likeness to the seed, plant, bud, flower cycle. Egg and seed are the beginning point. The stem and leaves, the actual leaf realm of the plant, and the caterpillar phase of the butterfly, represent the most vitality-filled phase in this metamorphosis. The real realm of growth in the plant is its leaf-realm. Everyone knows how clearly mowed grass and pruned trees express this vitality. It is just as well known how quickly little caterpillars grow large by living upon and nourishing themselves with the same green leaf. Without difficulty, chrysalis and bud can be characterized as moments of apparent standstill and stagnation. More or less in secret, the next phase is being prepared. When the flower and butterfly finally appear, what was in preparation becomes visible. That butterfly and flower express in their appearance an intimate connection, especially in their colorful splendor, is clearly seen.

Rudolf Steiner spoke of this connection in a lecture that he ended with a verse:

Behold the plant,
it is the butterfly
fettered to the earth.

Behold the butterfly,
it is the plant
*freed by the cosmos**

*Rudolf Steiner, *Harmony of the Creative Word.* (Rudolf Steiner Press, 2001).

The frog's [ex]change of form is a metamorphosis in the higher animal kingdom, the vertebrates. It is remarkable that here is still another [ex]change of form in outward stages, yet no explanation needs to be given. We can immediately understand that the mentioned [ex]changes of form, in such clearly separate phases, are only special examples of those that, in the end, take place in every animal. From the moment the animal starts to develop, we see an uninterrupted [ex]change of form—metamorphosis—that finally results in the mature animal form. Comparing this mature form to that of the egg, we can speak of a polarity. Thereby we can speak of the embryonic development as a continual *Steigerung* (progression to ever higher stages).

Superficially, we can compare the animal and plant metamorphosis. Yet it will be clear to many that there is a profound difference between the two. With plant metamorphosis we can, generally speaking, study the developed plant in its totality. The metamorphosis that takes place in the forms of the leaves, the sepals, the petals, etc., can be clearly seen on the entire plant.

This takes place very differently with animals. In the metamorphosis of the animal, each phase passes over entirely into the next. In other words, when the caterpillar becomes chrysalis, the caterpillar is gone; when the chrysalis becomes butterfly, the chrysalis is gone. Therefore, we feel we are dealing with an entirely different world. The point now is to characterize as clearly as possible the difference, in this respect, between the world of the plant and that of the animal. The best way to do this is again to ask the question: What is metamorphosing itself in the plant? Goethe said of the plant that it is a leaf from beginning to end. The metamorphosis of the plant can be read from the shape of its leaves, and we can say of the plant that it is the metamorphosis of a form. The metamorphosis is expressed in the sequence of its leaf forms—when observed from bottom to top. The plant is a form metamorphosis, as we have seen.

This does not apply to the animal. There is no question of the chrysalis being a metamorphosis of the caterpillar form, or the butterfly of the chrysalis form. This, in addition to the fact that each phase of an animal metamorphosis disappears into the next phase, can give us the feeling that we have to search for a totally different element. We have seen that the plant's metamorphosis comes about through outside influences—namely, the sun. This is why the plant is described as a being "between heaven and earth."

In the metamorphosis of the animal we must also search for that which is metamorphosing itself, and that is clearly something other than shape. For me this already lies hidden in the expression "exchange of form" (the Dutch *gedaanteverwisseling*). It would not be suitable to use this word for plant metamorphosis. In the animal there is something that is continually changing its appearance. What is this "something"?

The animal metamorphosis takes place between egg cell and mature form. Herein is already expressed its difference from the plant. The flower of the plant is the leaf principle which is brought to this form through the environment. One can almost say the flower not only belongs to the plant, it also belongs to the sun.

The animal metamorphosis takes place much more in the inner world of the animal. Our "something" works in the inner being of the animal, and when we ask ourselves what that "something" is, we come to the surprising conclusion that it lets itself be seen in the mature form. What then works as that "something" in the animal and continually takes on another appearance? It is that which we see as the mature form—the snail, the snake, the lion, etc. We may describe the animal as the result of an urge working within it which becomes visible in the mature form. How must we comprehend this urge? To do this we have to see the animal in its own environment.

Just as the plant lives between heaven and earth, the animal lives between itself and the environment. This animal life can be studied by everyone and can only be spoken of as a desire.

Two things, then, become clear. First, the word "desire" is not vague, and is not something that takes on the same form in every animal. The animal desire, not to be separated from the concept of the animal instinct, is something specific in each type of animal. When we also ask ourselves what exists in every desire, the answer as far as I can see can only be a striving for satisfaction.

Furthermore, when we question what sort of environment an animal lives in, then the answer is: in the environment that can satisfy its desire. We must keep in mind that in this connection environment is not identical with place. An ant, a butterfly and a lion have in the same spot of wilderness a totally different environment, depending upon their perceptions.

Thus, the metamorphosis of the animal is the metamorphosis of a desire that unfolds itself in its development from egg cell to mature form. The animal is a visible desire. My opinion is that the conclusion which can here be drawn, although perhaps surprising, is obvious: The animal is a unity of desire, form, and environment.

We can say that everything an animal does is a search for something in its environment to satisfy its desire. This leads us to the conclusion that, through the environment, the animal is continually brought to certain patterns of behavior.

Plant and animal metamorphoses have one thing in common: there where the flower brings forth the seed, the butterfly its egg, the mammal its egg cell, the cycle is closed; it takes place before our eyes on the earth. To make a long story short: plant and animal always stay "here."

Specific metamorphosis of the human being

How does metamorphosis work in human beings? In speaking about plants and animals, we referred to the principles that metamorphose themselves by the concepts *form* and *desire.* Concerning the metamorphosis of the human being, we must now search for the principle that distinguishes itself from what is at work in plant and animal. We immediately meet with a difficulty because the general opinion is that human and animal are not essentially different from each other—in short, that the human being is the highest mammal (in any case, an animal).

It is beyond the scope of this work to philosophize too much about this subject. Only in the last century did this idea begin to find acceptance in large circles; yet its roots began much earlier. In recent years many writers have been occupied with this subject and have tried to show the network of illusions around the idea of the human being as an animal.* A short consideration of this might set us on the right path.

Let us ask ourselves the difference between a mother expecting a baby and a dog expecting puppies. The difference becomes apparent only when the puppies and the baby are born. It could in both cases be said that what was born was what was expected. Yet now a remarkable difference starts to appear. One can ask: What will the little puppy become? The only right answer is: A big dog. What is the answer to the question: What will the baby become? If one should answer, "a grown-up person," one must take into account that in the mother there lives something more. When she holds the child in her arms, she will all too often ask: What will this baby become? It is clear the answer "a grown up person" will not satisfy her. She means something totally different, something she is still expecting.

When does this "something" that is being expected make its appearance? Superficially, each of us can have part of the answer by remembering what happens when one meets a young man or woman last seen as a baby. In a conversation with the proud mother one will often hear: Who would have ever dreamed!

However, we certainly cannot accept that the expectation has really been fulfilled. This would mean that one would expect nothing further from such a

*See Hermann Poppelbaum, *Man and Animal: Their Essential Difference* (London: Anthroposophical Publishing Company [Rudolf Steiner Press], n.d.).

young man or woman. We all immediately feel how absurd this would be.

We can go still further and ask how it would sound if we should say to someone over seventy with white hair: I don't expect anything from you anymore. Would a healthy person ever be satisfied with such a feeling about himself? I admit that we can only state this for healthy people; however, we have indeed found something in the human being that one can define with the statement: The real human in us is something that is always being expected.

Thinking this through, we can make a connection with our starting point: a mother is already expecting before the birth. The image of birth as a portal through which a being from out of the "here-before" (the Dutch *het hiervoormaals*) is expected, is not far-fetched, provided one has no prejudices against this idea.

Could not one also say the real human in us is expected through the portal of death—just as the real human was expected from that world by the mother through the portal of birth? The thought then presses upon us that what is essentially human, the metamorphosis of which we are seeking, is distinguished from the metamorphosis in the kingdoms of plant and animal because the latter always stay *here.* As has been seen, this does not apply to the human being; thus, the human is a being that during its life resides as a temporary guest on the earth.

Since the rest of this book deals with the problem of the metamorphosis of the human being, it may suffice to characterize the essence of human metamorphosis in a few sentences. In relation to the fact that humans and animals exhibit a great kinship with each other, we recall the concept [ex]change of form, and a question arises: Where in the human being do we find the real [ex]change of form? This can be found if the human being can be seen in the light of a certain law, to which the modern scientific world (somewhat understandably) provided a great resistance. This is the law of reincarnation.

If we consider the different forms in which an individuality lives on earth as the various forms of a metamorphosis, then here again we must try to find the law of polarity and intensification. To what degree this is possible shall later become evident.

It will be clear that this could never apply to animals. An animal specimen cannot be compared with a human being. An animal is born as a species already present, and remains so after the death of the specimen.

One could immediately remark that the human as species exists on earth before his birth and after his death. The important point is, however, that we find in the human being what is specifically human, what is above and beyond mankind as "species." This occurs when we deal with the core of the personality, with the individuality. In anatomy and physiology, and also in psychology, we deal with the human being as species. When confronted with a fact in anatomy or physiology it would never occur to one to ask: Whom are we speaking of here? In the study of psychology too we are dealing with soul qualities that, in essence, belong to every person.

When we are concerned with a biography, however, its contents apply only to one person in the whole world. When we note (especially in the higher animals) the differences within one species that might tempt us to speak of the biography of an animal, we meet a paradox. This paradox can be solved if one schools oneself to distinguish the differences between individuals from those differences which can only result from an individuality.*

Reincarnation can then be seen, as has been said, as metamorphosis, as [ex]changes of form of an individuality or spiritual being that in the realm of time and space reveals itself anew, again and again, in a biography.

This is what it means that the human being (in contrast with plants and animals) withdraws from the earth for a certain time at a particular moment in his metamorphosis. For now, this short indication must suffice. In Chapter 4 more will be said about this subject.**

Shape and being

Polarity and intensification have become well known concepts. But as we have said, they are phenomena that point directly to a unifying being which manifests itself in these phenomena.

We live in an age when, especially in biology, the word "being" (that invisible something which works creatively in the shapes around us) is no longer used; on the contrary, it is firmly dismissed as unscientific.

The only way out of this impasse is to discover, and to admit, that at the moment one passes from a single observation to a series of combinative observations, one discovers that the combination reveals more than the sum total of all the separate facts. Thus, one cannot remain in the realm of exact science. Whoever combines facts creates a composition and begins to see nature itself as a composition. This lands him irrevocably in the field of art. That is one of the reasons why there are so many opinions and disagreements in the fields of biology and anthropology, with regard to the doctrine of evolution. One can agree about the facts themselves; it is the composition that

*See Rudolf Steiner, *Theosophy* (Anthroposophic Press, 1994).
**In *The Dressed Angel* (Regency Press, 1975). I have tried to show that the idea of reincarnation can be responsibly spoken about only if one has first learned to discern that principle in the human being for which it would be thinkable and acceptable, as well as that to which it definitely does not apply.

can be interpreted in different ways and that makes it impossible to separate the scientific from the artistic. This leads to discord.

The resistance that an artistic viewpoint generally encounters in science has, to my mind, a very special background. The word "composition" evokes an association with the word "composer," and this introduces the notion of a creative being into the scientific discussion. This, in turn, introduces a religious element into our thoughts.

Here we have an important point of conflict between two groups of people who are engaged in the doctrine of evolution. One group reasons as follows: When we see a watch, we cannot help drawing the conclusion that a watchmaker has been at work. There is no watch without a being to create it. Because nature shows shapes that cannot be supposed to have originated from a haphazard coagulation of matter, we can only suppose that a creative being is behind the origin of the shapes of nature.

The adversaries of this theory agree that the watch has been made by a watchmaker, but they refuse to extend this conclusion logically to the shapes of nature because, according to them, it is not scientific to work with the notion of a watchmaker when he does appear as a visible being.

The slogan is: "No watchmaker." It can also be expressed this way: One wants to avoid at all costs a concept that makes one think, even a little, of religion, and whereby the whole creation would have to be considered to be the result of the activities of an unseen being or beings. It is easy to see that this difficulty can only be avoided by restricting oneself in "pure" science to the sensory observations without describing their context. That way one remains exact. But this kind of science would interest very few people.

Why is it that one insists on dismissing the idea of an unseen world as unscientific? We must not forget that this attitude did not emerge until a specific moment in the nineteenth century, when people came to accept only explanations from the visible world (Du Bois Reymond and Brucke in 1842). However, with this decision a dogma was created. To me, dogma is very unscientific.

When one notices that with the introduction of a scientific approach into a scientific discussion there appears on the horizon a religious element, one should not recoil from it. One should say: Knowledge of surrounding nature can give us complete satisfaction only when it appeals to the artistic *and* religious elements living in all of us. It will be clear that this has nothing to do with religion in the traditional sense.

Louis Bolk, in his theory of evolution, gave the opinion that the human being does not accidentally follow the animal realm but that the concept of humanity has played a leading role in animal evolution from the beginning:

> On the basis of adaptation and selection, evolution can only be a result, whereas I myself would rather view it as something fundamental. This point of view, when carried to its extreme, must lead to the conclusion that already in the lowest, let us call it primeval matter, lay ensconced the necessity of the "becoming of humanity." One might say that there would be a gradual buildup of new forms, each an improvement of the previous one, due to a factor that resides in the organism and regulates and controls the evolution. That is right, but I do not avoid the consequences. It may sound mystical, but I derive more satisfaction from it than from the consistent application of the theory of selection and adaptation as sole guiding principle of evolution, which would make the human being's appearance a mere fluke, a matter of chance.*

I believe that we, too, have gradually reached the stage where we must no longer avoid such consequences, but we must be careful not to introduce the previously mentioned element as a *hypothesis*. It is, in the truest sense of the word, the scientifically sound result of a certain approach.

Shape and genetic structure

In the part of science devoted to thoughts about the origin of shapes (for example, theories of evolution) there is, as we have seen, a great resistance to the notion of a creative being. People usually reject the existence of another reality working along with active forces in the visible world. The origin of shapes in nature are explained as a result of the genetic structure in their chromosomes. This is not the place to enter into philosophical discussions on the subject. A single practical example will illustrate my personal point of view.

It is often stated that our thoughts are the product of our brain structure. Our brain thinks. In order to prove this, it is agreed that a disturbance of the structure causes a disturbance in the thought process. When a musician plays the piano, its mechanical condition will influence his rendition of the music; failure in the instrument will result in a poor rendition. Yet nobody would say that the piano plays the music. *We* play, and to play we need a sound instrument. *We* think, and for thinking we need a sound brain.

It is the same in the relationship between a creative being and the so-called genetic structure. If this structure is disturbed, one will see a misshapen manifestation. When will people understand that the many variations on the

*Louis Bolk, *Hersenen en cultuur (Brain and Culture)* (Amsterdam: Scheltema en Holkema, 1918), p. 49.

Drosophila fly, obtained by a forcible outside influence on the genetic structure, are nothing but the misshapen manifestations of its essential being?

If we then question the origin of the genetic structure, it is not difficult to find an answer. It is the result of the activity of the creative being that has formed a shape down to its finest details, even down to the genetic "formula."*

Trying to explain the origin of a form by its genetic structure only pushes the problem further away, where it remains just as unsolved as before.

We experience the same pattern with each creation of an artist and in each product of our technology. This is not apparent to us because the creative, spiritual human being, whom we all know from our own experience, is more and more misunderstood and overlooked.

The notion that life, soul, and spirit are only functions of material forms has nevertheless a very special background. It is connected with the way in which the concepts of matter have developed over the centuries in our cultural life. These concepts unavoidably shook the previously existing conviction that the visible world had been created by a spiritual being. For if one considers life, soul, and spirit as realities that lead their own existence in our material world, one must be able to understand the origin of the shapes in which life, soul, and spirit appear.

In the past this presented no problem. A real spiritual world that cannot be seen by us as such, but that reveals itself to us in nature around us, was considered self-evident. This changed when human beings endeavored to visualize the essence of matter. This endeavor led to the question: Can the old notion that the world has been created by a spiritual being (God) still be maintained? How can a spiritual being with totally different qualities from those we attribute to matter interact with matter?

In order to understand this we must first be aware of the deep gulf that exists between the two concepts "spirit has shaped matter" on the one hand, and "spirit is a function of the material shape" on the other. This I have discussed at length in my book *The Dressed Angel.* There I tried to demonstrate that shapes never function, that it is always the creative principle itself which functions with or in the shape it has created. Besides this, the most essential question for us now is: How can we bridge the gap between spirit and matter?

The origin of shapes

Here we are dealing with two things: that which shapes, which expresses itself, and that which we call the material, in the sense of an artist's material.

*Formula means "little form."

An artist is a tangible, visible being, and we have no difficulty in accepting the fact that he is capable of creating shapes out of matter. When it comes to shapes in nature, and therefore in the skeleton, we shall again have to accept a creative principle, as we have called it several times, even though we are not dealing with a visible being. It is the same principle we discussed in the previous chapter in connection with the idea of metamorphosis.

Here arises an understandable question: If we must characterize a creative being as nonmaterial, invisible, and transcendental, how can it touch and handle concrete matter as we know it? To put it bluntly, how can spirit grapple with matter?

This question really concerns only the *origin* of the living shapes in evolution. Once they have appeared on earth they are governed by the well-known law: Living things bear living things. We are always dealing with living things on earth, as much in hidden life (the seed or the egg) as in revealed life (the flower or the animal).

Perhaps it is necessary to give a definition of the concept, "life." I will restrict myself to a quotation from Professor Bolk's book, *Brain and Culture*: "To me life is a formative principle sui generis." Life and the creation of shapes are thus made identical. I completely agree with this.

To understand the original synthesis, "life-matter," one must remember that matter here means the dead matter from the mineral realm. Furthermore, one must remember that the mineral realm is composed of four elements—earth, water, air and warmth—a differentiation well known in former times, gradually forgotten, and replaced by the notion of aggregate conditions. It will be a surprise to many to be told that these four elements also show the principle of metamorphosis, up to a point. It must be understood that metamorphosis in this instance has, of necessity, a different meaning, because in the mineral world the word "shape" has a different essence. Apart from the crystals, which are another question, everything here is amorphous, formless. Quantities, qualities, and characteristics are the essence here.

To find a connection with metamorphoses we must try to see that in the mineral world, too, some things can appear in more than one guise, as in the shapes of living nature. Again, too, we will have to find polarity and *Steigerung*.

We must try to see that earth, water, air, and warmth are different manifestations of a common phenomenon. It is known that air, water, and earth can change into each other's condition. Innumerable materials are known in a gaseous, liquid, and solid state. But until a short time ago there appeared in physics an unabridgeable gap between warmth and matter. Heat was considered as the kinetic energy. A transition, for example, from warmth to air, was unthinkable in a scientific context.

All this has changed since the conclusion was reached that matter—solid, liquid, and gaseous—must in the last resort be seen as condensed energy. This removed the essential difference between warmth and matter. After all, kinetic energy and energy mean the same thing.

May this suffice for the moment. Energy, therefore, is the principle we were looking for when we set ourselves the task of visualizing the four elements as each other's metamorphosis. Energy shows itself here in four forms: warmth, air, water, and earth. Obviously we are dealing simultaneously with a polarity and an intensification (*Steigerung*) of condensation or dilution. Thus we bridge the gap between being and matter, if the latter is thought of as originating through the condensation of a condition of warmth.

What is energy? Where does this word come from? In human life, to have energy means to be capable of activity. This concept is closely related to the word *will.*

In the preceding pages we have often spoken of shaping *forces.* The word "force" is employed repeatedly in chemistry and physics. But the word "force" is a completely empty, abstract concept unless we think of a being from whom that force emanates. I know only too well how far modern scientific thinking has drifted away from that conclusion. Yet at the same time, I am equally convinced that it stands quite close to it. Years ago a friend of mine said that will is a force capable of overcoming an obstacle. That is a very good description. Will is here described as a force.

If this will to create existed "in the beginning," and we ask again, "How could spirit grapple with matter?" we must visualize this matter as warmth. Warmth is the bridge, the link between will and our condensed matter.

Thus we obtain a strange, even miraculous, sequence: willing, being, force, warmth, matter, shape. Speaking of shape, it is obvious that these factors can bring forth a shape only if the concept "viewpoint," spoken of earlier, is again brought to mind. Without this there can be no question of a content or idea that must be at the basis of each shape, unless it is to remain only a contour.

It is everyone's daily experience that a spiritual being is a source of will, strength, and activity; therefore, from this viewpoint this being is capable of creating shapes. Although at present the human is the only visible creative being on earth, he is not any the less a spiritual being.

In olden times the world was viewed as a creation of spiritual beings. This vision, this form of knowledge has gradually been lost. Anthroposophy discusses these things, called hierarchical beings, in the same concrete terms we employ for the visible beings around us.*

We are still far from understanding the transition from being and energy to condensation of warmth into visible shapes. The latter are manifested, moreover, in many species, such as plants, animals, and humans. However, in scientific thought the concept "energy," which turns out to be the essence of matter, is equally far removed from the final state in which it appears as the mineral realm.

I am conscious of the fact that I have only given the merest hint of a solution. The only conclusion of the preceding is that in a living being we are not dealing with matter as molecular construction. The image of this construction applies only to so-called "dead matter."

For now, a great many questions will have to remain unanswered. My principle aim was to show how the molecular ideas about matter that have developed over several centuries inevitably lead to the contemporary materialistic outlook on life. A molecule is a matter in a condition, and we cannot imagine how "spirit" has "seized" it.

Yet, to my mind, this development is extremely important, because one has been forced to think along sharply formulated lines. But to remain consistent, this thinking (which is constantly improving its powers of discrimination) must be ready again and again to renounce accepted theses and adjust deeply seated ideas.

This has, indeed, been happening. But every now and then there is the tendency to say: "*Now*, at last, we do know." We must not allow our thinking to become rigid; we must keep it supple. Let us remain receptive to new discoveries and ideas.

*See, e.g., Rudolf Steiner, *An Outline of Esoteric Science* (Anthroposophic Press, 1997).

Chapter 3
Metamorphosis in the human skeleton

We will first examine metamorphoses in the skeleton. We should never forget that we must test the result by the law we have promulgated for every metamorphosis. Even when we return later to the idea of reincarnation, we will have to prove if and to what extent we can speak there, too, of polarity and development (*Steigerung*).

The following illustrations show the metamorphoses of shapes of human bones in several ways.

In the beginning, one may feel directed toward a number of unrelated phenomena of which one cannot make head nor tail. This changes when one begins to distinguish not only shapes but also, with the aid of several examples, the parts of the body to which they belong.

The description will be restricted to explaining what each picture represents, which bones are involved, and so forth. For the rest, the illustrations must speak for themselves. Occasionally it will be indicated which detail should be studied with special care in order to grasp the essence.

The skeleton as a whole

The first illustrations show the whole skeleton. Plates 5 and 6 are shown to demonstrate that in the past man was never considered totally separate from his surroundings. One can see clearly how closely art and science were still linked in those days in the playfulness and expressive gestures of the skeletons and their surroundings. The vegetation that has undoubtedly been drawn on purpose beside the stone and rock formations seems to express the thought that the skeleton was not all that dead. There is even a cherub in one of the pictures.

Pl. 5. Drawing by Wandelaar from an anatomy textbook by Bernard Siegfried Albinus (1697-1770).

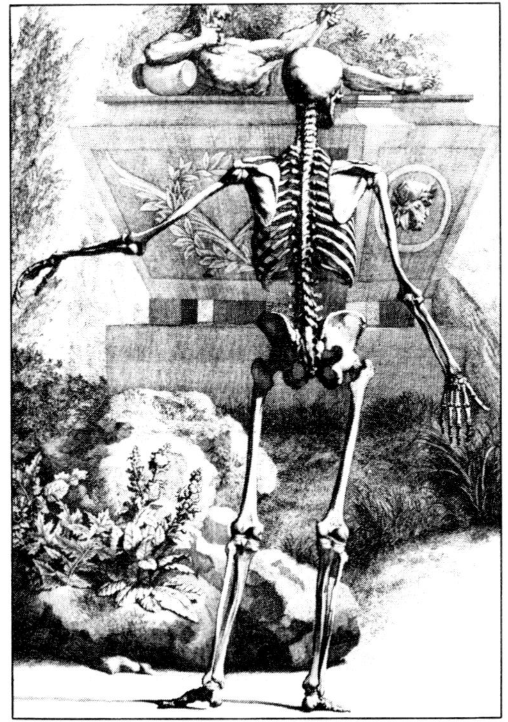

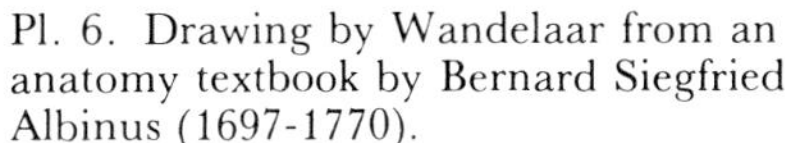

Pl. 6. Drawing by Wandelaar from an anatomy textbook by Bernard Siegfried Albinus (1697-1770).

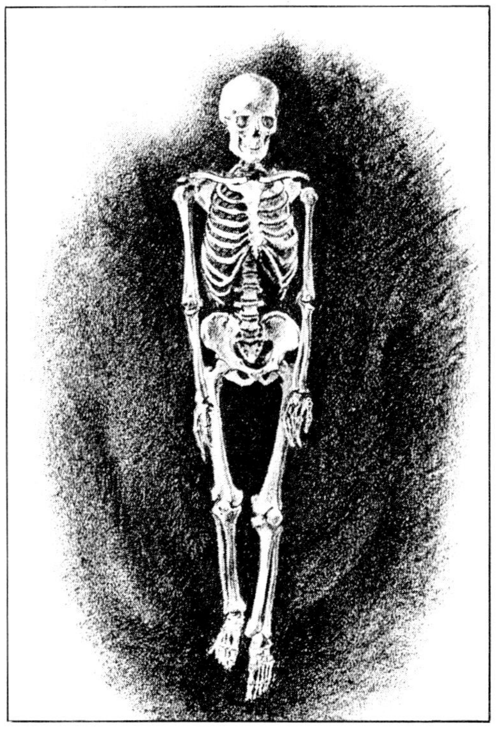

Pl. 7. Drawing of human skeleton by Rupert van der Linden.

Plate 7 is a drawing by Rupert van der Linden, who expresses the same idea in an entirely different way. Plates 8 and 9 are photographs of a section of the upper ends of the thighbones and the heel bones. They show the generally familiar structure of the lamellae, which shows in a remarkable way how our body (especially our skeleton) obeys laws of mechanics and statics. It may be significant for some people to discover here the beauty of static vectors.

Looking at the skeleton as a whole, one is immediately struck by the fact that one can actually distinguish three areas: the head; the thorax and shoulder girdle with the arms; and the pelvic girdle with the legs.

The head has a strikingly small, concentrated, rounded shape. Except for the lower jaw, the bones are fused together by seams called sutures (pls. 10 and 11). The skull is composed of the bones of the face with the lower jaw and the cranium. The latter looks like a dome; the former is characterized by the many openings connected with our sense organs. On the sides are the acoustic holes.

The thorax is notable for the rhythmic repetition of the shape of the ribs. The dozen pairs of ribs enclose the thorax and are attached to the sternum by

Pl. 8. Section of upper ends of thighbones.

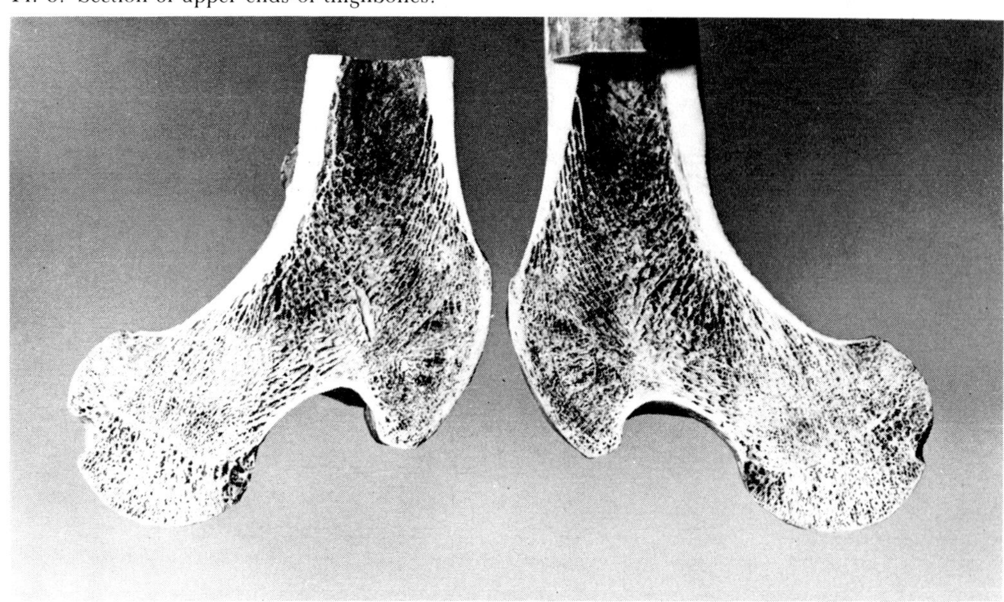

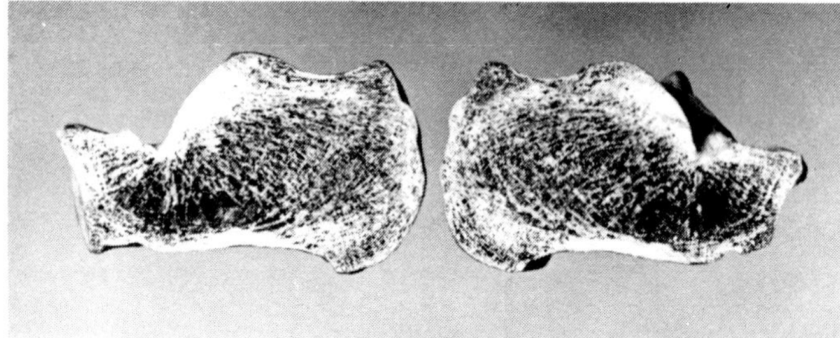

Pl. 9. Section of upper ends of heel bones.

Pl. 10. Cranial bone illustrating sutures.

Pl. 11. Cranial bone illustrating sutures.

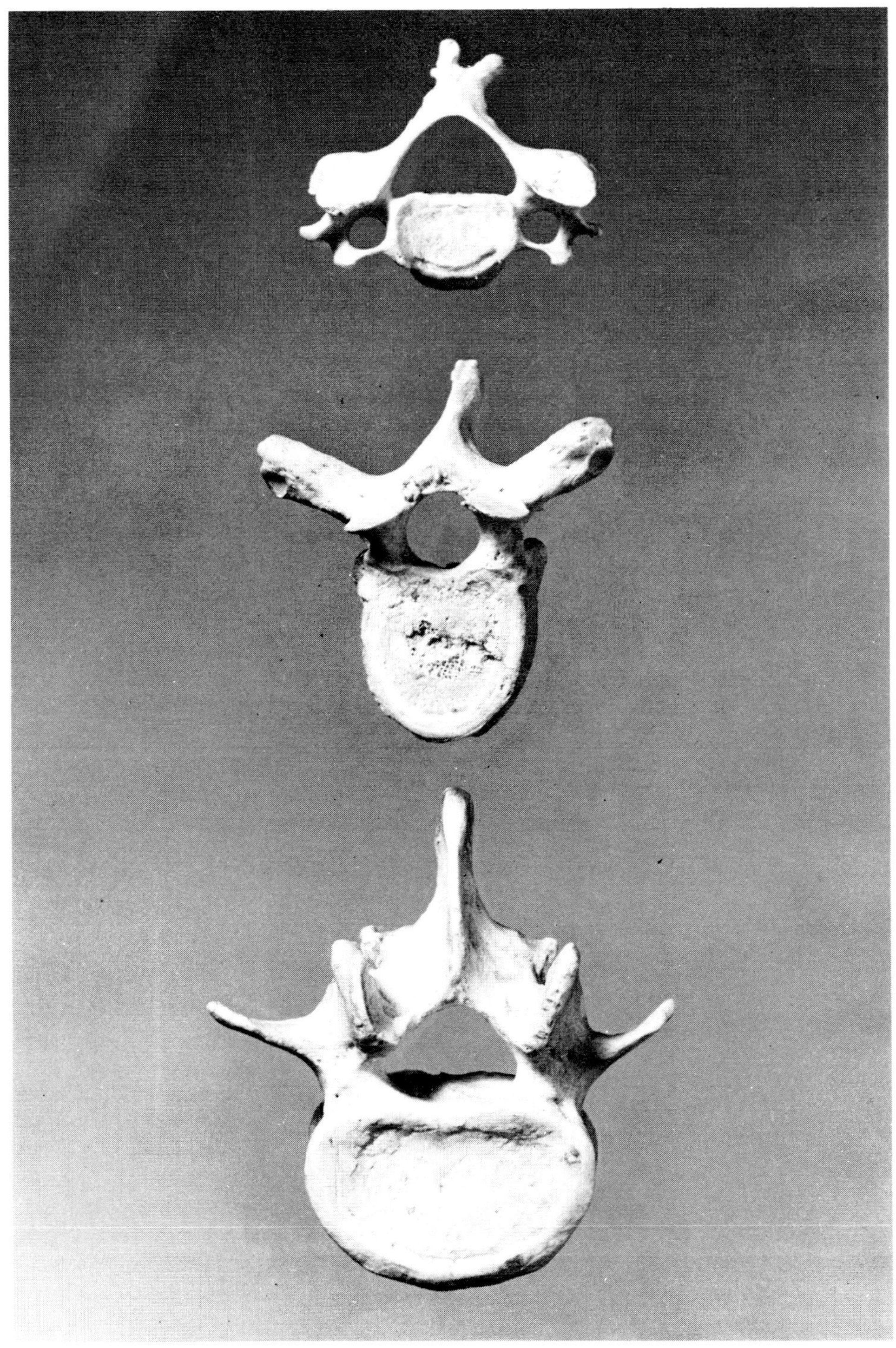

Pl. 12. From top to bottom: cervical vertebra, dorsal vertebra, lumbar vertebra.

cartilage except for the two lowest, the so-called floating ribs. They are attached to the spine by joints at the back.

The shoulder girdle consists of shoulder blades and collar bones. The shoulder blades are actually fusions of originally separate bones. The comb on the shoulder blade is remarkable.

The pelvis consists of two hipbones and the sacrum. The latter is a fusion of the five sacral vertebrae. The arms form joints with the shoulder blades. In the same way the legs join with the hipbones on the spot where three formerly separate bones have fused into the hipbone.

The spine runs dorsally through the whole trunk. It is a very important part of our skeleton and is composed of various types of vertebrae. The vertebrae have a typical structure: a solid, large oval centrum; the body, to which is attached dorsally the vertebral arch, with the dorsal processes; and transverse processes extending laterally from the arch (pl. 13). Arch and body enclose the vertebral canal.

Like most mammals, humans have seven cervical vertebrae of a fine structure, with pierced transverse processes; twelve thoracic vertebrae that support the ribs; five lumbar vertebrae, heavier and coarser than the others; and five sacral vertebrae that are fused into one sacrum. Several caudal vertebrae are fused to form the coccyx.

The spinal column is a splendid example of metamorphosis in our body. One can see a few examples in plate 12. From top to bottom they represent a cervical vertebra, a dorsal vertebra, and a lumbar vertebra. In plate 14 we see the first cervical vertebra, the atlas.

If we take a look at the whole, we can describe it as the spinal column with thorax, on the one end supporting the head and on the other end standing on earth with the pelvis and the legs. Shoulder blades and arms are connected to the thorax only by the front of the collar bones—thus, hardly at all.

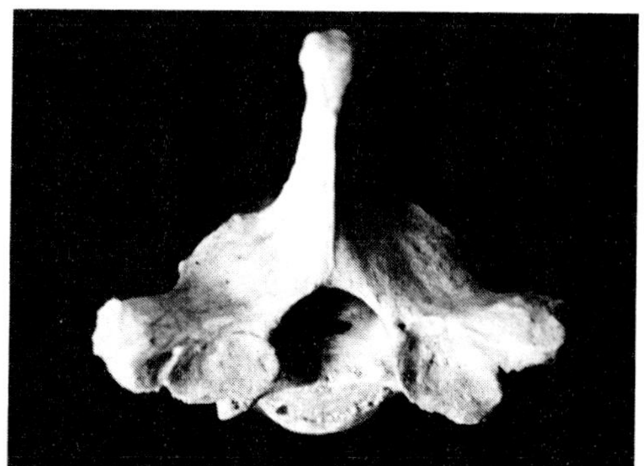

Pl. 13. A spinal vertebra.

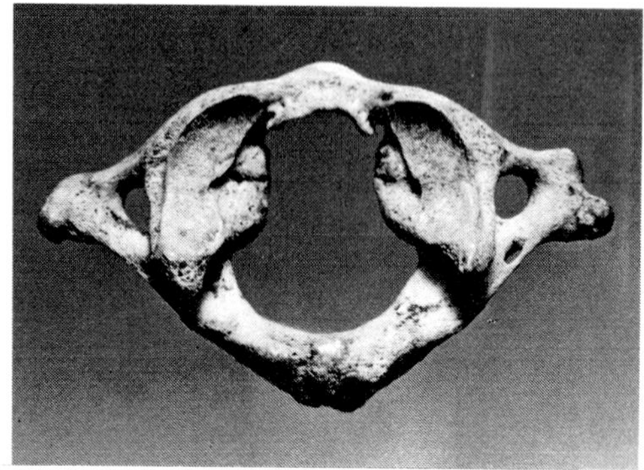

Pl. 14. The first cervical vertebra, the atlas.

P. 15. Parietal bone.

The bones of head and limbs are absolute polarities. The skull is round, enclosing and concentrated, consisting of flat or irregular bones that are joined by seams (except, of course, the lower jaw). By contrast, the limbs are linear; they consist of the so-called long bones. Later we will learn to discern here a radiating principle. The various parts are connected by movable joints. There could hardly be a greater contrast. Compare plate 15, which shows a parietal bone (part of the skull), with plate 16.

Since the human being is the only truly upright creature on earth, the

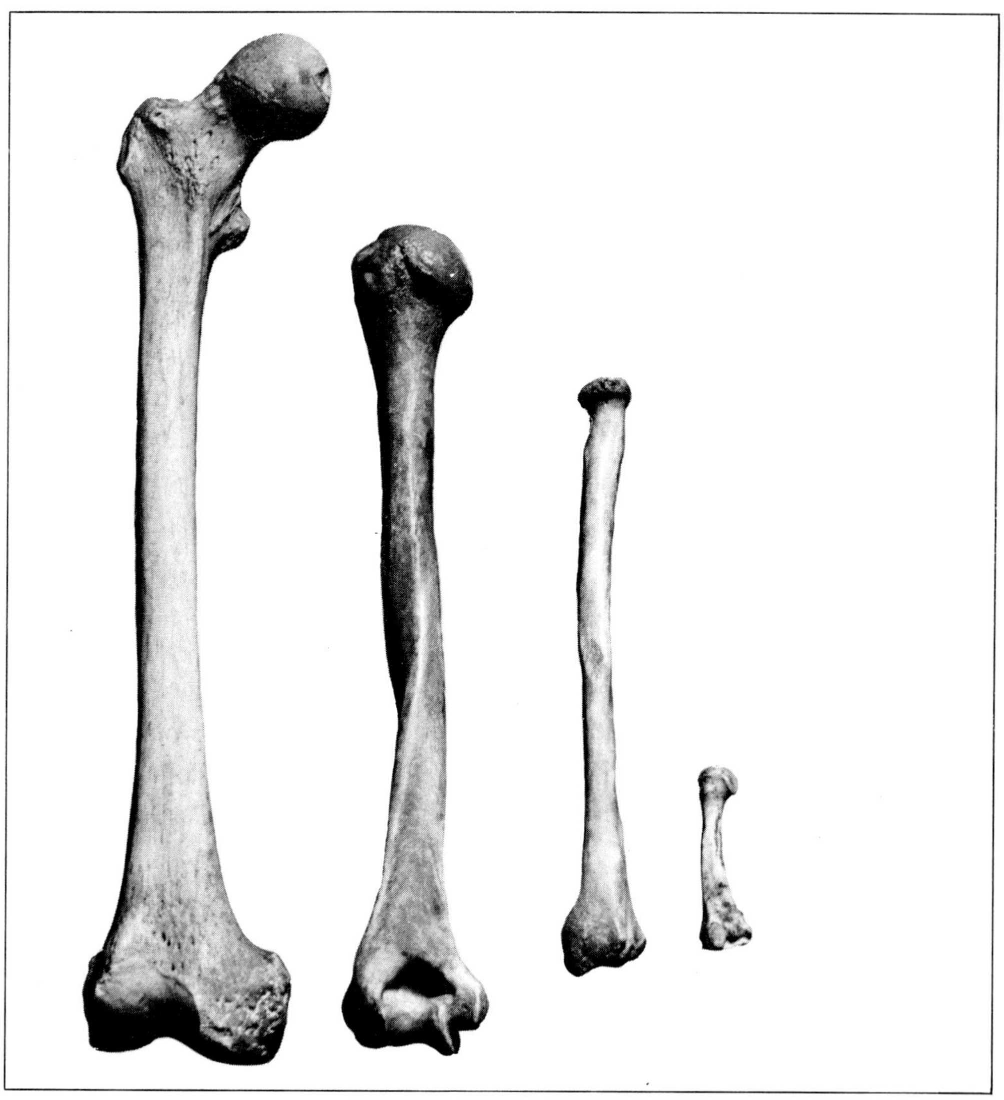

Pl. 16. From left to right: thighbone (femur), upper arm bone (humerus), lower arm bone, metatarsal bone.

arms occupy a special position because they are also in the service of our spiritual life—for example, in gesticulation. Hands folded in prayer vaguely imitate the shape of the skull. Between the two is the thorax, which in its construction has a bit of each: it is enclosing, but less closed than the skull, and much more mobile.

It is worth mentioning that the structure of the thorax is a rhythmic repetition of more or less similar shapes. In the contrast between the head as a pole of *rest* and the limbs as a pole of *movement*, we encounter a primeval law:

Wherever in nature rest and movement meet, rhythm appears. This is seen, for example, in the vibrating string when touched with a bow, the rustling leaves of a tree when moved by the wind, the ripples in sand caused by wind and water.

This last example will help us understand that we can recognize the same phenomenon in our body if vertebrae and ribs are compared with ripples on a beach. We may imagine that they are the result of engendering forces that were active in the past. These examples show how an action in time can become a phenomenon in space. Such an image can become a window through which we can observe the once creative activity and, in our imagination, call it back to life.

The thorax is composed of three areas: uppermost is the fixed part; the first rib is directly attached to the breastbone without cartilaginous mediation; underneath are the mobile, floating ribs. In between are the typical ribs that are attached to the breastbone by cartilage. Cartilage guarantees a certain mobility.

The frontal skull can also be divided into three areas: forehead, nasal, and mouth and jaws. The first is a dome; the second shows inside the rhythmic repetition of the nose shells (conchae). Of the two jaws the lower is mobile; both represent the limbs of the head.

With a little imagination, the same principle can be seen in the arm and leg with their threefold structure of upper arm—upper leg, lower arm—lower leg, and hand—foot. Here we think especially of the remarkable ball and socket joint of the shoulder and hip. The lower arms and legs each have two bones; the hands and feet are most mobile.

It is not difficult to find the same idea in the hand, which is divided into the carpus, the metacarpus, and the fingers.

Going a step further we look at a particular bone, the thighbone (pl. 17), and can, after observation, see the outline of a rather hidden but complete human form: a head, a neck, a pair of shoulders, a slim torso, even a pair of knees.

Years ago, I met a doctor who had noticed this and who illustrated it in front of a small audience. Kneeling down with his arms crossed and his head a little to one side, he endeavored to imitate a thighbone. Not long ago a friend of mine sent me reproductions of works by the Belgian sculptor Minne, who expressed in his creations exactly what I have tried to say here (pl. 18).

Visiting an art exhibition by pupils of the Waldorf school in Nienstedten near Hamburg, I saw drawings made by children in the ninth grade who had just been learning about the skeleton. One of the drawings is reproduced here

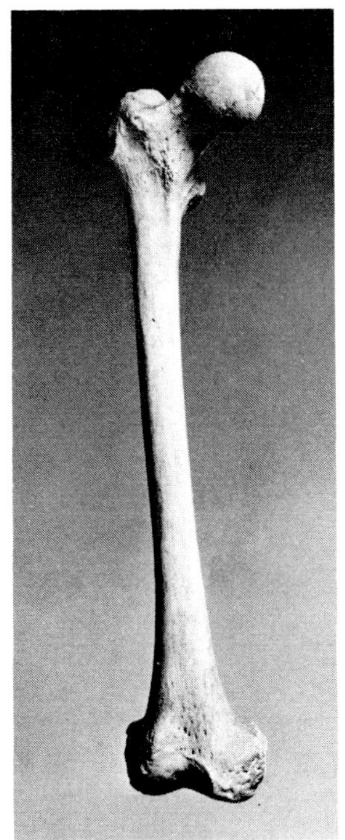

Pl. 17. Thighbone.

Pl. 18. George Minne, *Kneeling Boys*, fountain in Ghent, Belgium, 1898.

(pl. 19). The teacher assured me that nothing discussed here had even been mentioned in class. He himself never heard of it. I eagerly appropriated the drawing; I could not have wished for a better illustration of what I have said above. For the sake of completeness I have inserted one more statue by Minne and a thighbone, both in the same position as in the drawing (pls. 21 and 20).

To be able to understand the total metamorphosis of this form within hidden places, one finds on plate 16 a photograph of a thighbone, upper arm bone, lower arm bone, and metatarsal bone. The metatarsal bone by itself would not suggest the idea of the seed of the whole human form. However, when one studies this series of four shapes and allow one's eyes to wander from left to right and back, then the outer resemblance can mean the beginning of the experience of an inner affinity. Here one can gradually learn to know metamorphosis, to feel it and to experience it.

If we consider a rib to be a special kind of long bone and we look at the part that twice forms a joint with the vertebra, we can distinguish here also a head, a neck and a few knots that again evoke the image we have just dis-

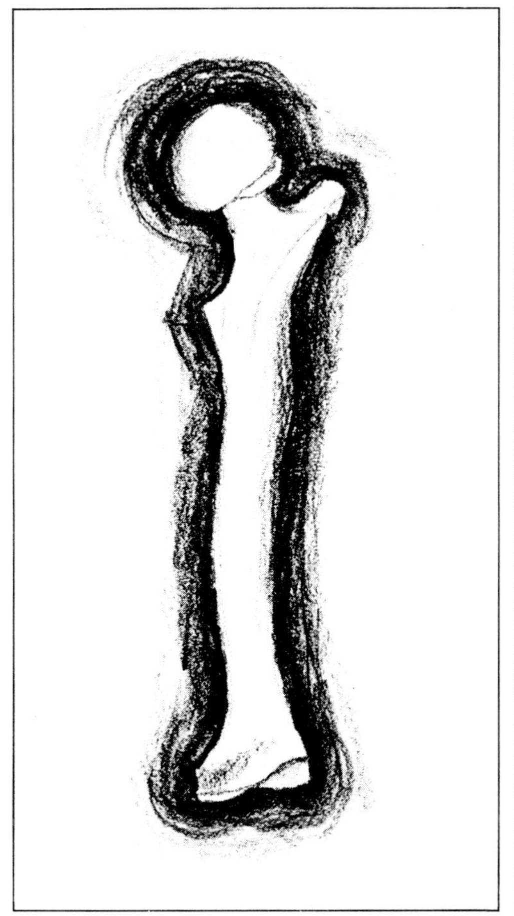

Pl. 19. Drawing of thighbone by ninth grade student in Waldorf School in Nienstedten, West Germany.

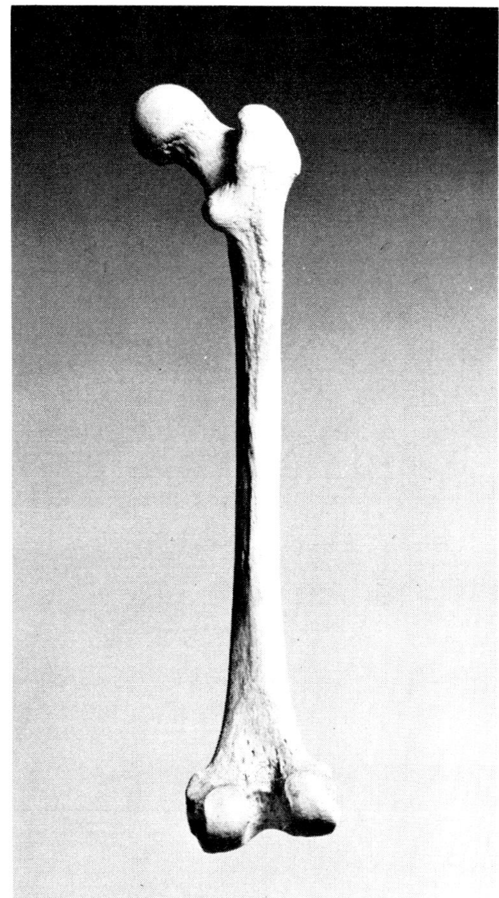

Pl. 20. Thighbone.

cussed (pl. 22). The only difference is that the rib does not have a loose end; its extremity plunges into a cartilaginous mass or ends in itself (floating ribs), so there is no image of lower limbs. One could visualize a dozen shrouded human forms, enclosing the thorax on both sides, without a shaped end. We are reminded of two dozen mermaids.

Thoracic vertebra and shoulder blade

It is fairly easy to recognize at a glance a certain resemblance between the arch of a thoracic vertebra and the shoulder blade, both in dorsal view (pls. 23, 24, and 25). It is true that at first sight many people will raise questions, as when one notices that the corresponding elements of shape have totally dif-

Pl. 21. George Minne, *Kneeling Boy*, 1898.

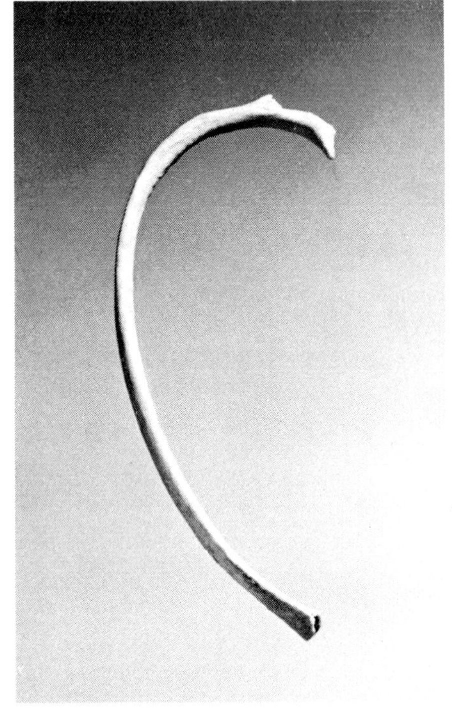

Pl. 22. A rib.

ferent functions. The outward curve at the top of the vertebra provides a surface for joining with the next vertebra. No such thing can be found in the shoulder blade. This teaches us, however, to distinguish between form and function. I stress this because in contemporary books and lectures on natural science, function always takes first place. The question arises as to whether this resemblance is coincidental or far-fetched. With every new comparison this question will arise again. It is only when one notices that all these so-called far-fetched resemblances begin to form a striking whole that one will come to see how the far-fetched is also close at hand.

The reason for observing only the back of vertebra and shoulder blade lies in the fact that we find in our body—and elsewhere in nature—areas in which concepts such as strength, mass, and activity are prominent, while

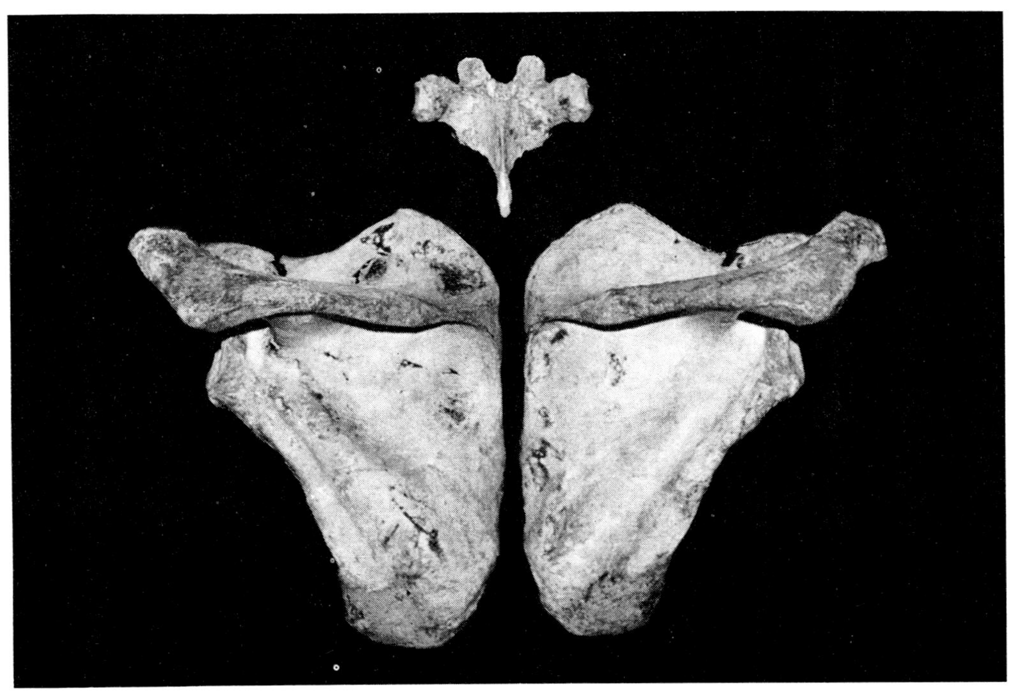

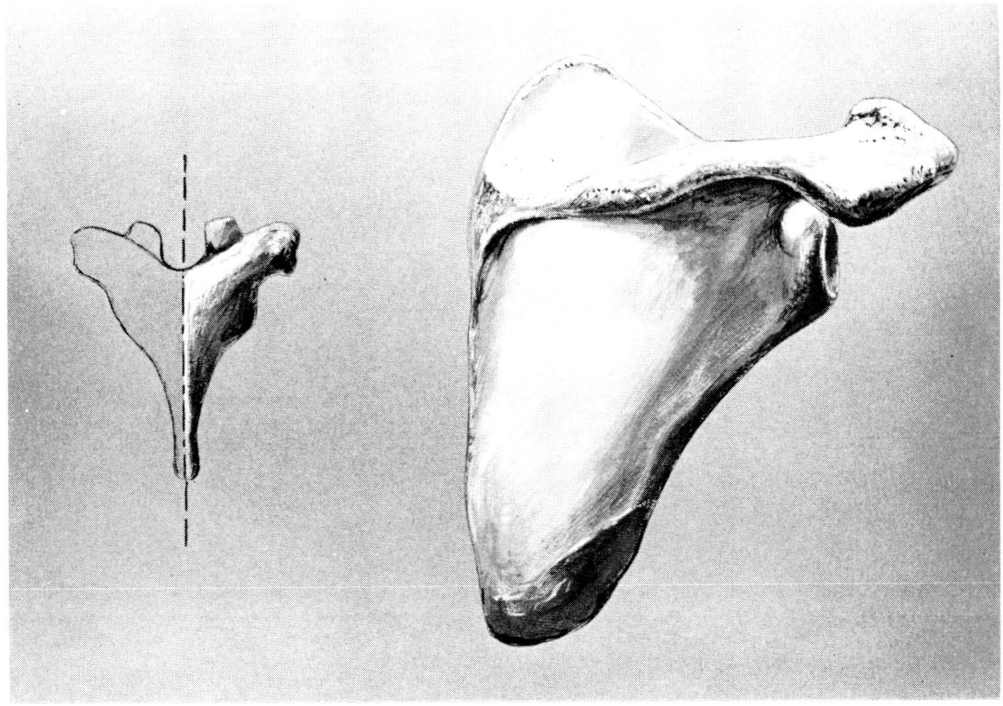

Pl. 24. Thoracic vertebra and shoulder blade.

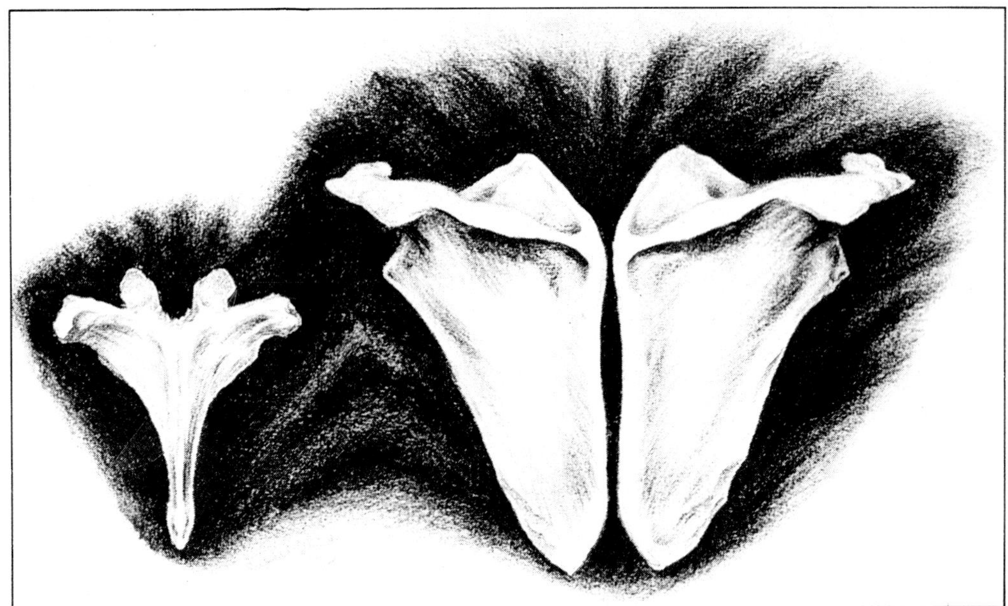

Pl. 25. Thoracic vertebra and shoulder blades.

other areas confront us more with the concepts of image, expression, and rest.

A vertebra offers a typical example. The front shows the so-called body, a more or less oval shape. The bodies together form the column that supports our spine. One might call the dorsal side of the vertebral arch the "face" of the vertebra. Therefore one could say that the shoulder blade has no body at all. It is only face. In anatomy, too, the word is *facies* (Latin for face).

It is too early at this stage to speak of metamorphosis. Up to this point, we are dealing with only an exterior resemblance of shape. We must examine the following examples of similar resemblances before we can arrive at the conclusion reached for plant metamorphosis. To achieve this we must not only observe the exterior affinity of shape but remember that each thoracic vertebra is articulated with a pair of ribs, just as the shoulder blades are with two arms.

Ribs and arms

We have seen that the twelve ribs enclose the thorax. The arms, in contrast, are loose limbs. The twelve thoracic vertebrae and the ribs form a unit of twelve small elements. The shoulder blades and arms form a corresponding unit.

What is the mathematical structure of the arm? We have one upper arm bone, two lower arm bones, and the carpus (wrist), which from the back clearly shows one row of three and one row of four carpal bones. Then there is a row of five metacarpals, then the five fingers which radiate visibly outwards. This radiation can be seen as a progressive emanation that is clearly expressed in the series 1-2-3-4-5 (pl. 26).

Experts will say that the carpus has not seven, but eight, bones, the eighth being the pisiform which lies at the base of the ball of the little finger. This is true, but it is equally true that this bone has nothing to do with the joint formation in the above-mentioned series. It is, so to speak, stuck onto the carpus on the medial side of the front of the wrist, and thus does not contradict the principle of radiation mentioned here. The tarsus, moreover, has no eighth bone (pls. 27 and 28). For the sake of clarity we have printed a reproduction from Spalteholz's anatomy book in which the series 2-3-4-5 can clearly be seen (pl. 29).

How can we describe the ribs? We might call them twelve fixed rays that envelop the thorax. The arms can be seen as two big rays, not fixed but free in the service of a very different task, which also includes the gesture of embracing. The celebrated Oken, a contemporary of Goethe, expressed a similar impression: "The arms are ribs opened to the front, nothing new, only freed."

Plate 30 shows a picture of a vertebra with ribs; plates 31 and 32 show the shoulder blades with arms, seen from above. Plates 33 and 34 portray similar views. In plate 33 we see the whole thorax from above, with the shoulder blades, the collar bones, and the arms. In plate 34, we see shoulder blades and arms from above, and a vertebra with the upper ribs.

Shoulder blade and hipbone

Our next task is to see if something similar can be applied to the pelvic girdle. The hipbone, part of the pelvic girdle, does not show any direct resemblance to the dorsal aspect of a thoracic vertebra, as does the shoulder blade.

There is, on the other hand, a marked resemblance in shape between the hipbone and the shoulder blade, if they are juxtaposed in a special way (pls. 35, 36 and 37). By holding a left shoulder blade upside down next to a right hipbone, we discover the affinity of shape.

The shape of the shoulder blade shows, as we have seen, a similarity to the arch of a thoracic vertebra. It follows that the hipbone also must have a similarity of shape, although more hidden. This will be more meaningful if

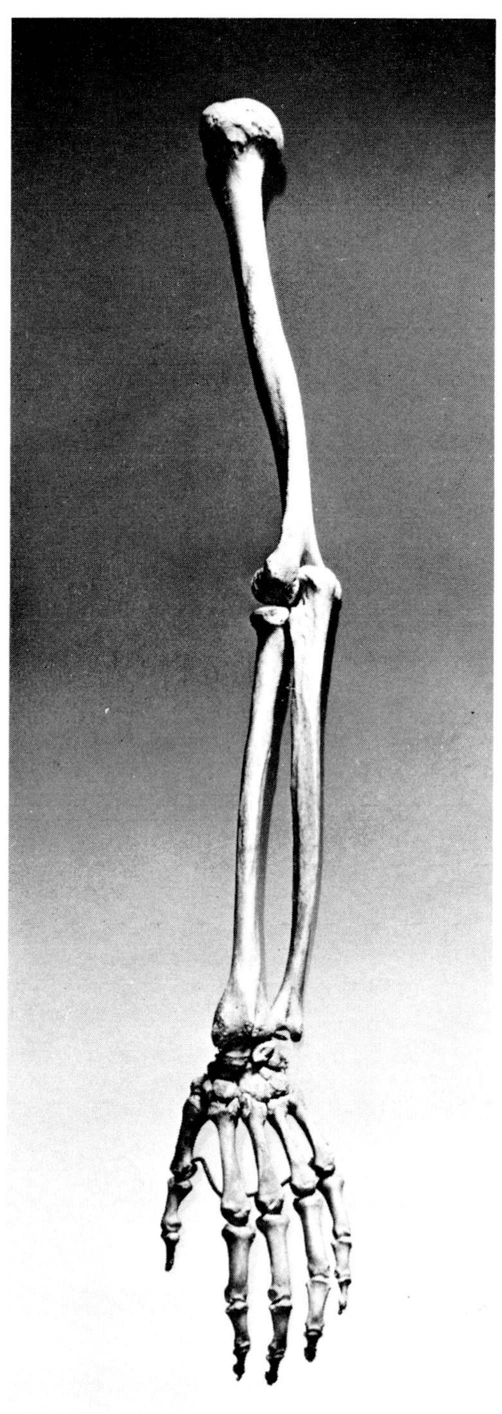

Pl. 26. Bones of arm and hand.

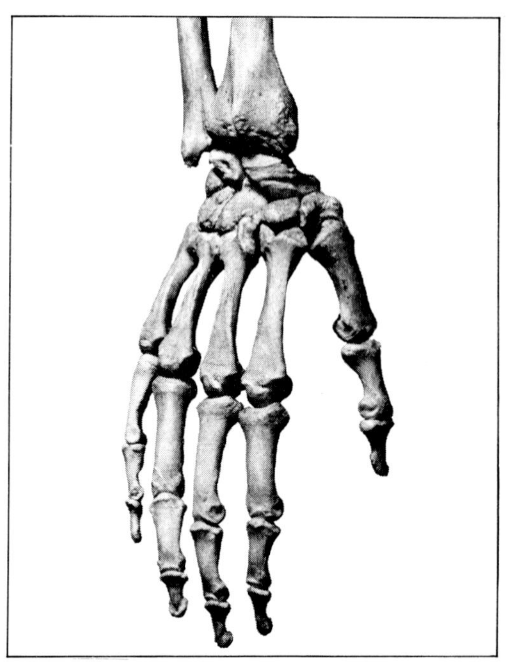

Pl. 27. Tarsal and hand bones, front view.

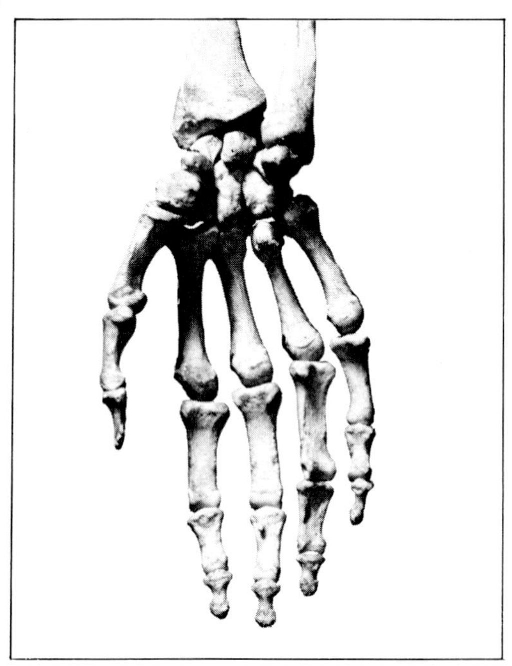

Pl. 28. Tarsal and hand bones, rear view.

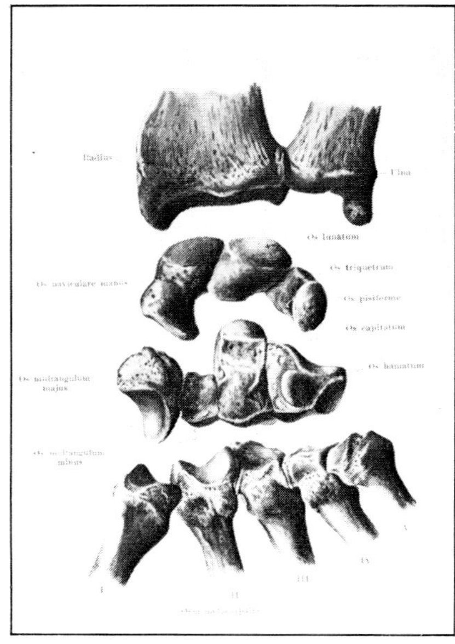

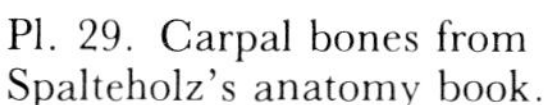
Pl. 29. Carpal bones from Spalteholz's anatomy book.

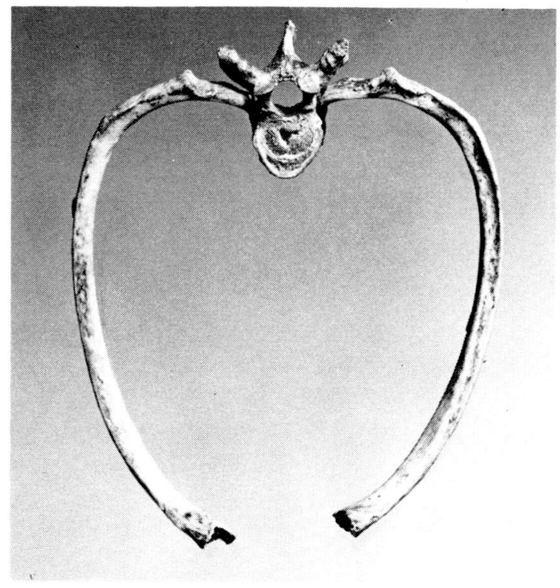
Pl. 30. Vertebra with ribs.

we remember that our shoulder blades are connected to our arms and our hipbones to our legs.

Ribs and legs

A leg is made up of an upper bone (femur), two lower bones (tibia and fibula), a tarsus (ankle) with a clearly visible group of three large and four small tarsal bones lying neatly in a row, five metatarsals, and five toes (pls. 38 and 39). The structure is, therefore, the same as in the arm: 1-2-3-4-5, and radiating, though in a totally different manner, because our feet do not move freely in space like our hands, but serve to walk on the ground. This is also the reason that the articulations between underleg and tarsus show different constructions from those we find between underarm and carpus.

Conclusion

It is already possible to arrive at a tentative conclusion: we have discovered a shape-theme—a vertebra with two ribs. This appears twelve times in the thorax and, in a different form, twice more in our girdles with limbs.

Another argument in favor of the radial principle of the limbs, which can be taken as a gesture of liberations, is the following. When we visualize the

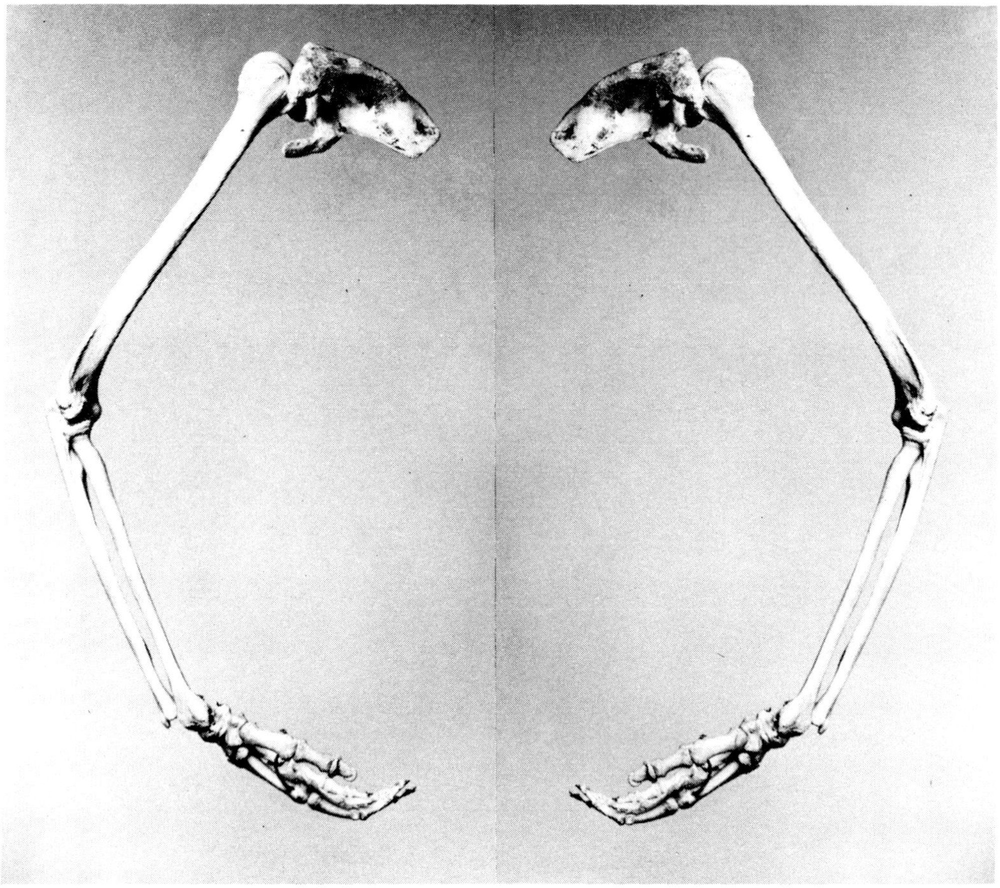

Pls. 31 & 32. Shoulder blades with arm bones.

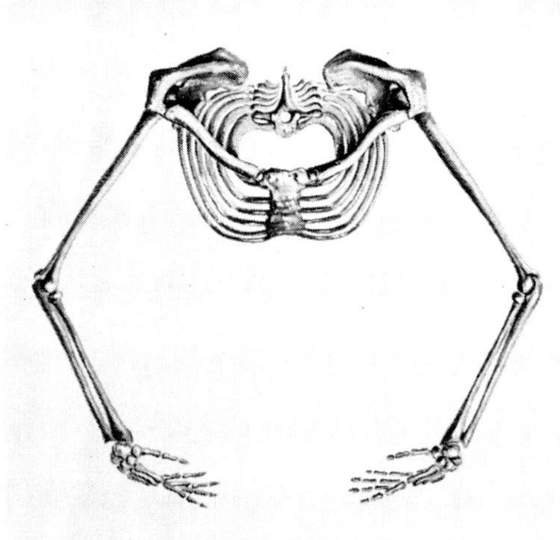

Pl. 33. Rib cage with clavicle, sternum, spine, and arm bones, seen from above.

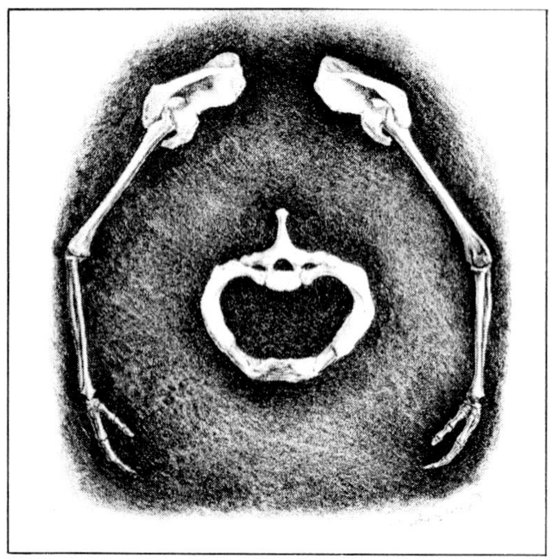

Pl. 34. Drawing of spine and rib cage with shoulder blades and arms, seen from above.

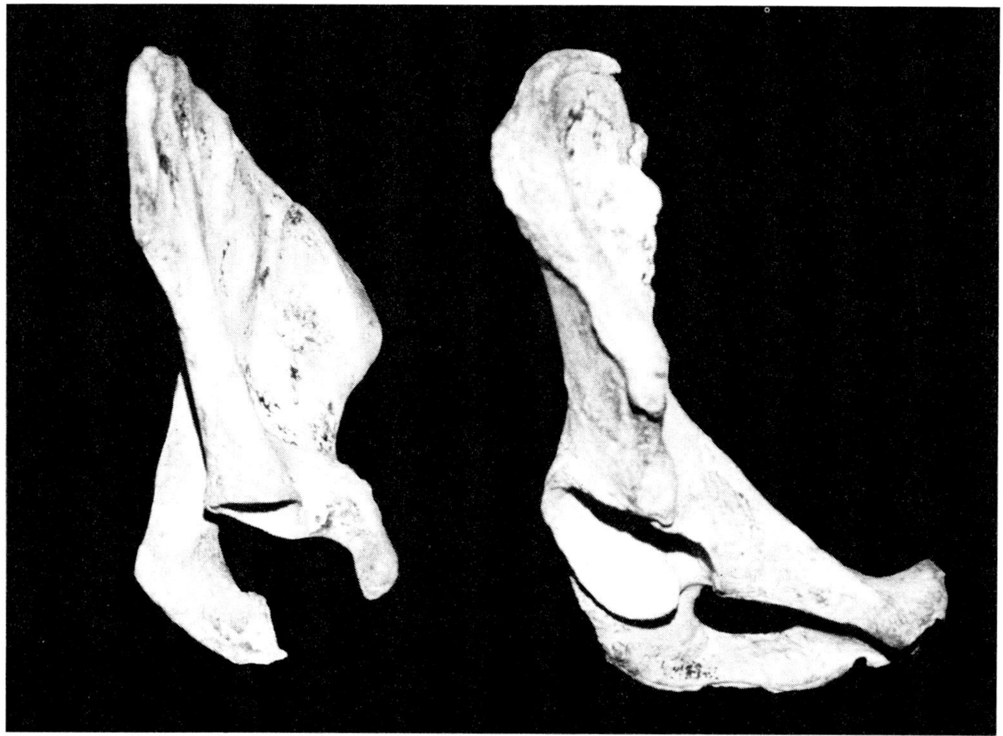

Pl. 35. Right hipbone and upside down left shoulder blade.

skeleton as a whole, the head, the dome, is directed upward as opposed to the downward-radiating limbs. We see the same phenomenon when we compare the whole human figure with the thigh bone.

In the upper arm bone and in the thighbone, the dome-shaped head is on top and the part comparable to the limbs is below. The same is true for the lower arm bones—the radius and the ulna—although perhaps less obviously so. In the lower leg bone the same applies to the tibia and the fibula. The "head" is directed toward the center, the "foot" towards the periphery.

This is followed in the hand by the carpus, in the foot by the tarsus. If we then look at the position of the metacarpals and the metatarsals, we see a complete reversal. The metacarpals and the metatarsals direct their "foot" toward the center and their heads clearly toward the periphery.

To see the metatarsal as it is in our body, it would have to be turned around in plate 16. This continues to the last phalanges of the fingers and toes. We have, therefore, a complete reversal in the carpus and the tarsus. This can clearly be seen in plates 27 and 39.

Now one can see that both the arm in the carpus and the foot in the tarsus orient themselves in a very special way toward the periphery. But "head"

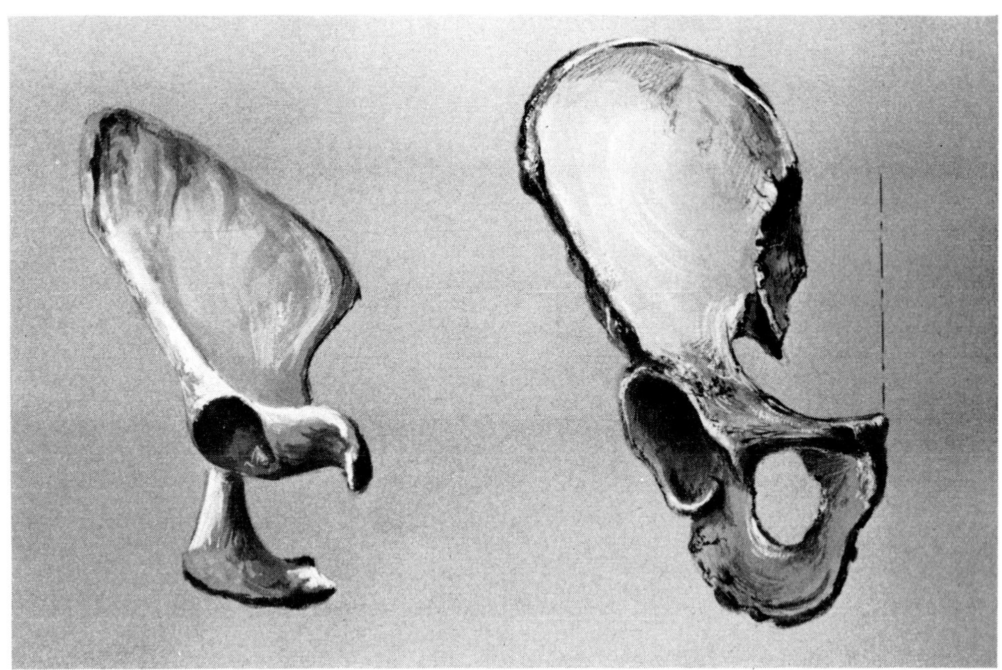

Pl. 36. Right hipbone and upside down left shoulder blade.

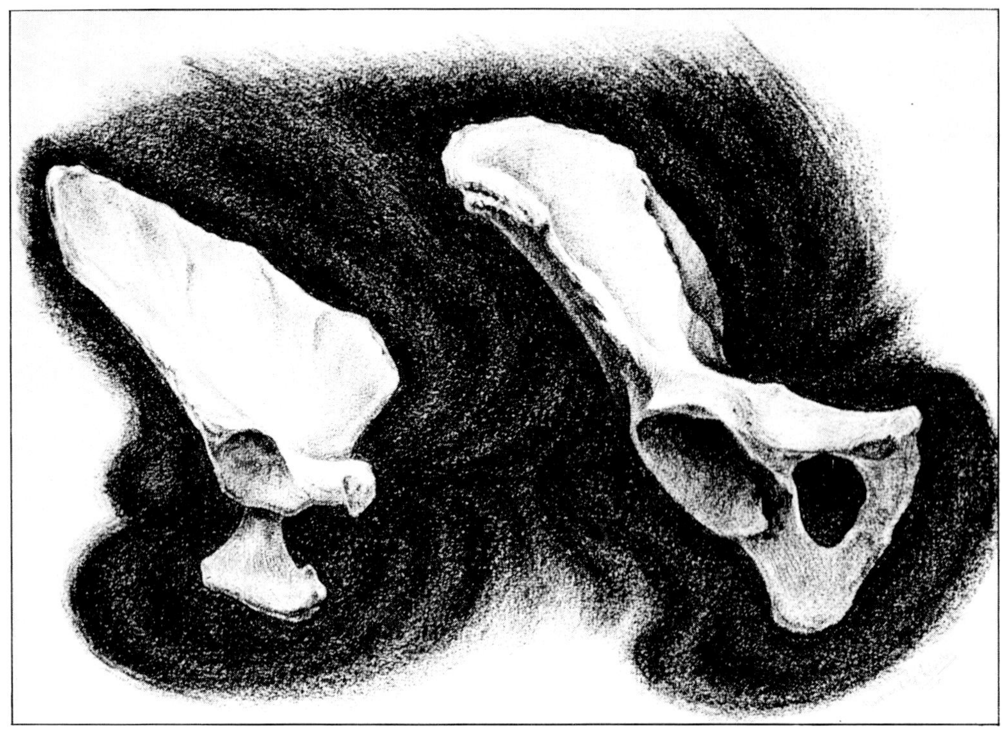

Pl. 37. Drawing of right hipbone and upside down left shoulder blade.

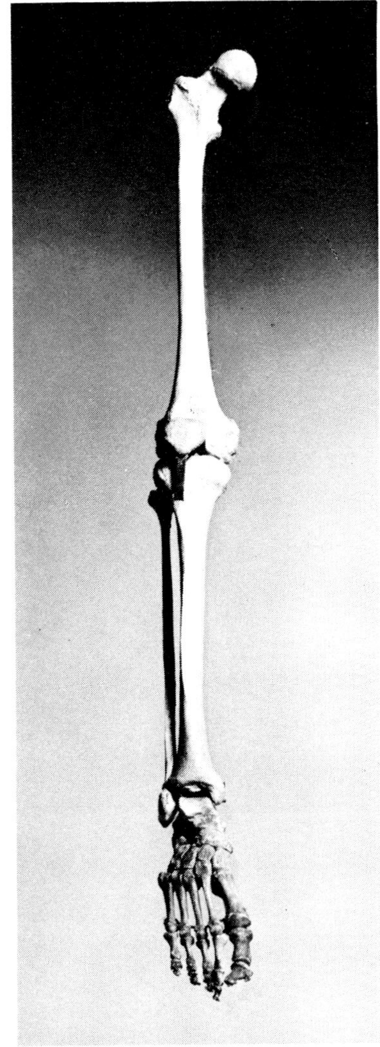

Pl. 38. Leg and foot bones.

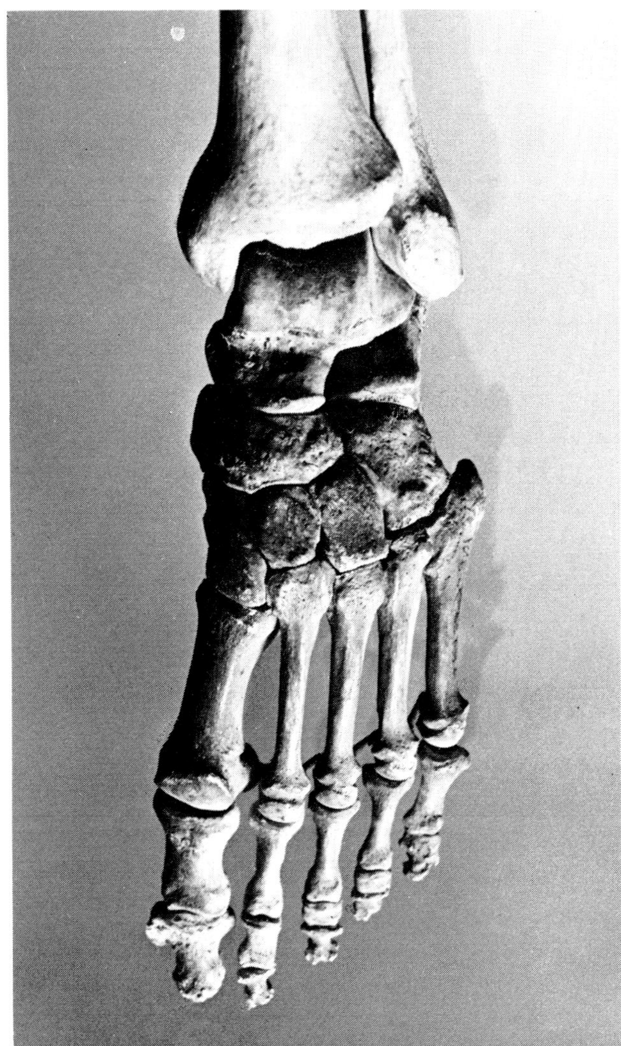

Pl. 39. Foot bones.

also expresses its connection with consciousness, with our observation pole. In our head it is the eye and the ear, as well as other sense organs that direct our attention to the outside world; in the limbs this is done by the fingers and toes. Our two hands have, so to speak, ten little "heads."

Rudolf Steiner once said to the teachers in the Waldorf School: "The kleptomaniac sees too much with his fingers." This function of the fingers is clearly visible in the structure of the skeleton. We see and feel everything with our fingers, but they are also the field of our action. As time passes we leave behind us an ever larger trail of actions. Thus we arrive at a threefold radiation:

1) a spreading in the total limb structure
2) observation of the outside world through touch
3) taking action in the outside world

What a contrast to the head with its watching, its concentration in observing, and its small, closed shape!

On the one side we have the moving joints of the limbs, and on the other the immovable joints of the bones of the skull (with the exception of the lower jaw). When we think of this contrast, we realize how differently oriented the forces must have been that have called it forth. This will be confirmed by later discoveries.

Lower jaw and thighbone

Now we are entering upon the second stage of our research and will make the first tentative connection between the shapes of the trunk skeleton and certain bones of the skull.

Years ago a friend of mine who was interested in the shape of bones showed me the similarity between the so-called rising branch of the lower jaw and the upper part of the thighbone, that is to say, the head, the thigh neck, and the appertaining thigh knobs.

This similarity will at first elude many. The issue is the dual image of the head and the opposite projection, with the bowl shaped concavity in between (pls. 40, 42 and 43). My only starting point was that a friend had seen it. These first impressions are often confirmed or corrected by handling the bones and becoming familiar with their shape.

When one compares the back of the upper part of the thighbone with the back inside of the equivalent part of the lower jaw, the mysterious connection between these shapes, strongest in the play of lines between the various protrusions, becomes even more apparent. In the beginning, we pointed to the possibility of recognizing a total human form in the thighbone, and this notion was strengthened by looking at the work of the sculptor Minne. In the same way it is possible to experience the play of gestures of the protrusions of the lower jaw (including the corner of the jaw), as well as the play of gestures occurring between the stumps of limbs (pl. 41).

When we remember that the whole leg consists of thigh, lower leg, and foot, and that we have compared the lower jaw only with the thighbone, the question arises as to whether this comparison, resemblance, or similarity of

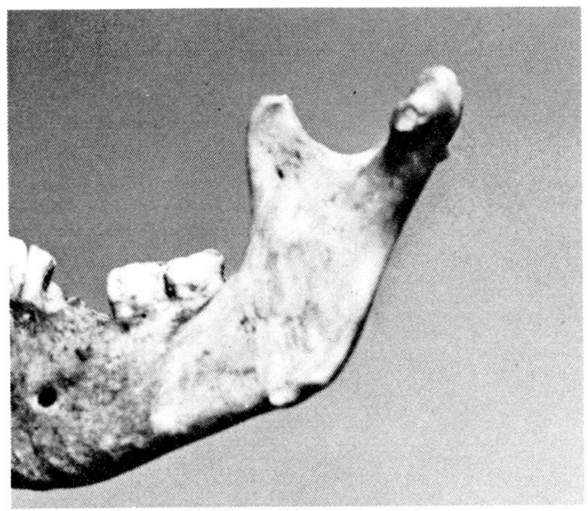
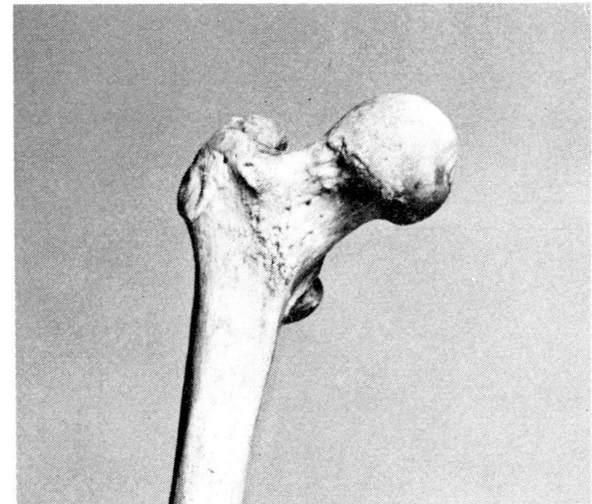

Pl. 40. Left: lower jawbone; right: upper end of thighbone.

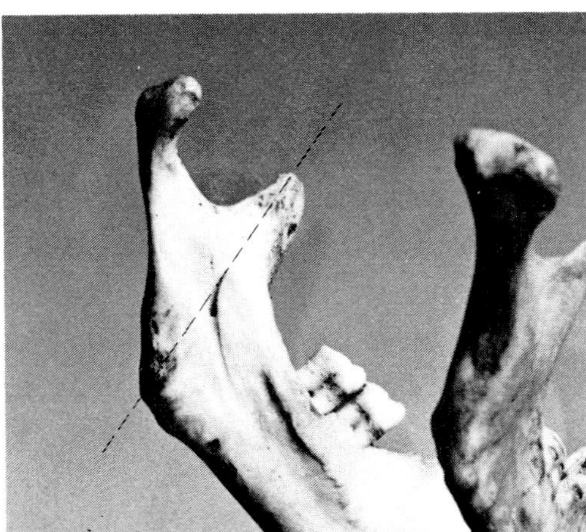
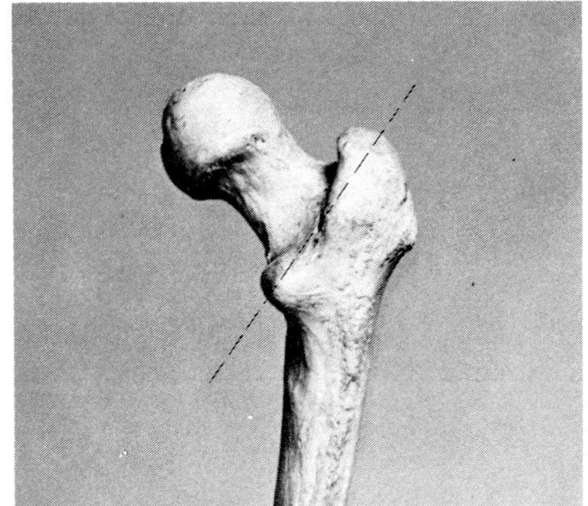

Pl. 41. Left: lower jawbone; right: upper end of thighbone.

shape holds for the rest of the leg. One will also have to recognize a certain threefold element in the lower jaw. This can certainly be found in:

1) the part between the joint and the corner of the jaw
2) the horizontal part from the corner up to the chin
3) the rising part of the chin up to the front row of teeth (pls. 44, 45 and 46).

I realize many people will find it difficult at first to accept this comparison, but by keeping an open mind one can in time view such a fused bone as the lower jaw in an entirely new light. To this end one should remember that the lower jaw ultimately represents a pair of totally fused limbs and is the only mobile element of the skull.

If we visualize the whole leg in such a way that the joints of the knee and

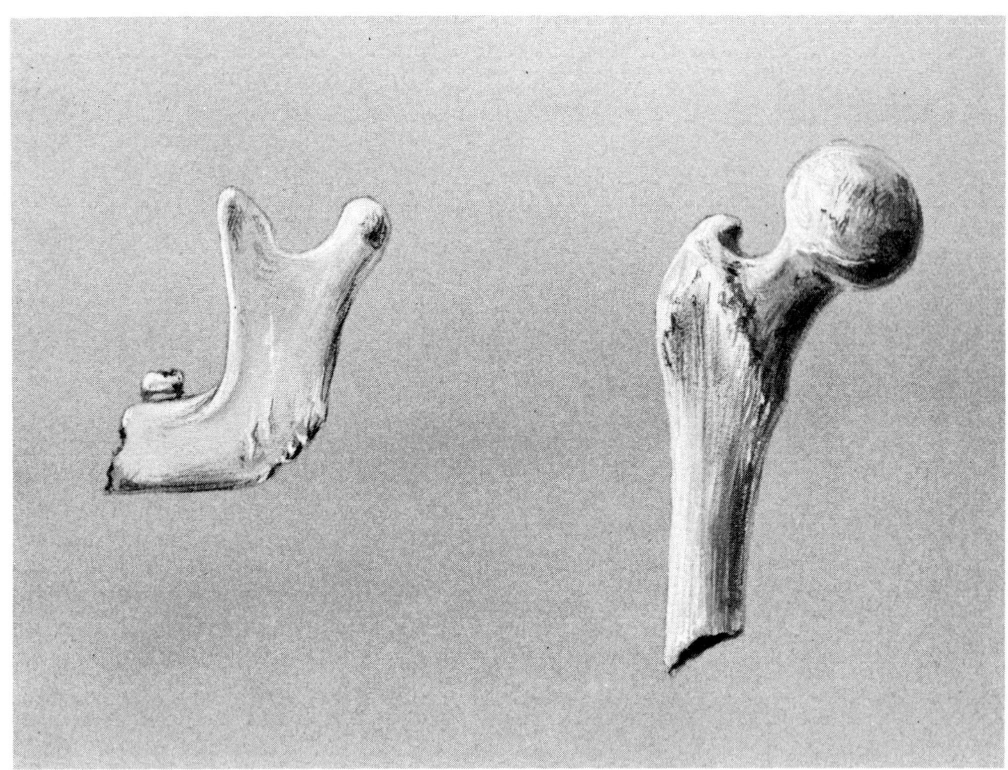

Pl. 42. Drawing of lower jawbone and upper end of thighbone.

Pl. 43. Drawing of lower jawbone and upper end of thighbone.

Pl. 44. Lower jawbone and leg bones.

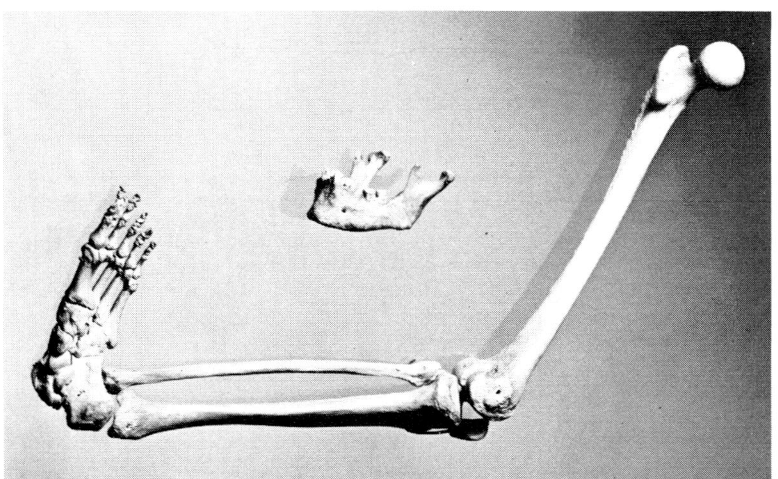

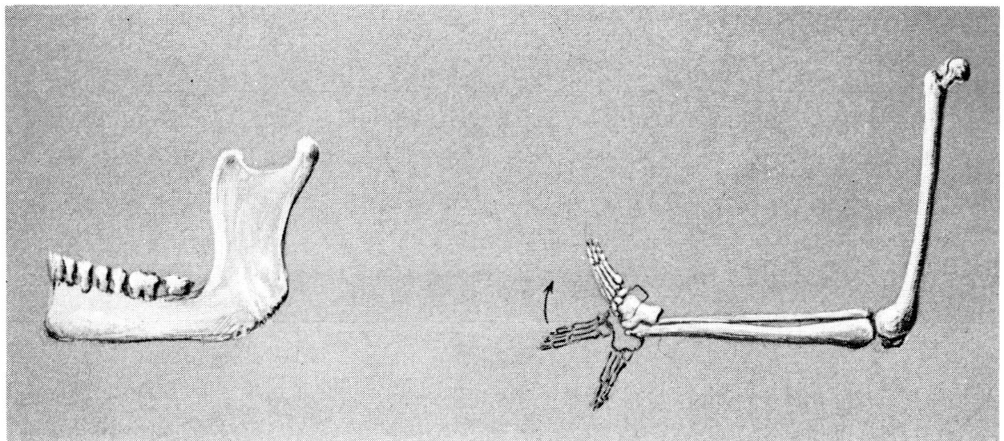

Pl. 45. Drawing of lower jawbone and leg bones.

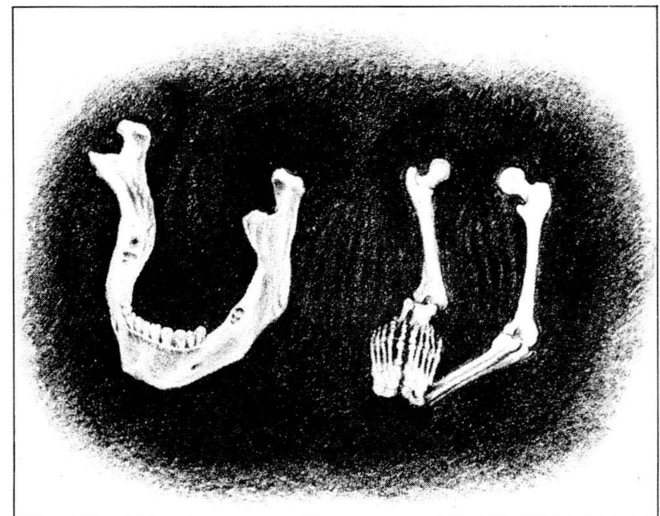

Pl. 46. Drawing of lower jawbone and right and left leg bones.

foot are bent in a position that imitates the shape of the lower jaw, we encounter an apparently insuperable difficulty: We cannot bend the foot in the same direction as the knee; moreover, the heel is in the way. We shall see later that the possibility of bending the foot this way can be dealt with in a manner not yet forseeable. The fact that the foot can eventually be bent backwards can be considered, but it is impossible to do away with the heel.

As a result of the fusion of two movable bones into the jawbone, we can see on the inside of the chin a small projection (pl. 47 from above, pl. 48 from below) where we would expect the heels. In usual studies of the skeleton this projection is seen only as the place where muscles are attached; that is to say, in connection with a function. Though this is plausible, it in no way contradicts the train of thought we have just suggested. It is only a separate approach. One and the same thing can have two different aspects.

It is certainly a generally accepted fact that the shape of the bones has been influenced by the use of muscles. How these shapes originated, how the muscles developed in our body, is another question altogether for which contemporary anthropology has no explanation. Here we have a clear example of the possibility of considering form and function as separate entities.

We are not denying, of course, that form and function are closely related. We must beware, however, of seeing form exclusively in relationship to function. This would land us in the principle of teleology, a concept that has been justly eliminated from natural science.

We are not trying to find an explanation for the human shape, but rather

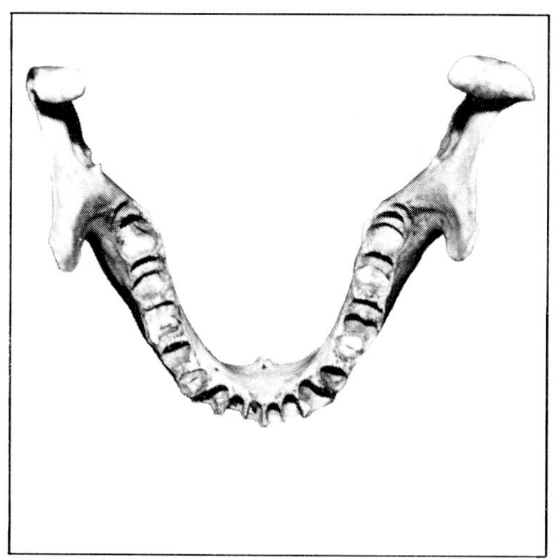

Pl. 47. Lower jawbone viewed from above.

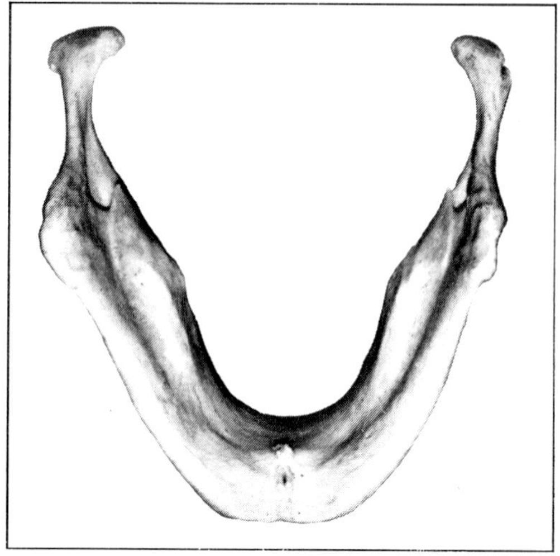

Pl. 48. Lower jawbone viewed from below.

to take it as a point of departure so we may, as far as possible, orient ourselves in the field of the creative forces.

Upper jaw and arms

Just as we have compared the lower jaw with the leg, we now turn our attention to the upper jaw. In spite of all appearances, our upper jaw, like our lower jaw, must be considered as a limb. Therefore the upper jaw must be seen as the equivalent of the arm. Although we were still able to trace a vague resemblance between the lower jaw and the leg, we find no such thing in the upper jaw. This is mainly because the upper jaw has been transformed into a different condition. The nasal and oral cavities and the eye sockets have altered the original shape almost beyond recognition.

Nevertheless, we must think of such a connection. Plates 49 and 50 show a skull and separately a maxillary bone, with a hand and part of the lower arm below. We can compare here only the part of the maxillary bone that holds the teeth with the hand and its fingers. Using a little imagination, the grooves on the upper jaw remind us of small long bones fused together. If we tie in the comparison between legs and mandible, we can point out ten fingers and ten toes, complete with nails. There are ten milk teeth in each jaw. For now we must restrict our comparison to the mention of a tenfold hardening in each jaw.

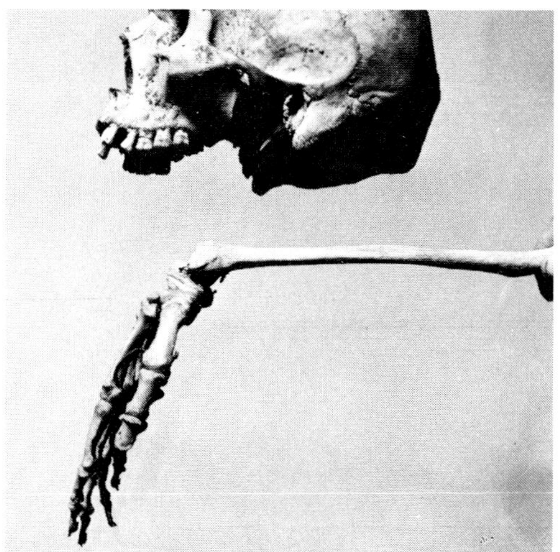

Pl. 49. Skull with bones of hand and lower arm.

Pl. 50. Maxillary bone (upper jawbone) with bones of hand and lower arm.

Hipbone and temporal bone

The mandible has been compared to the leg. First we will look at that to which our legs and our mandibles are attached—the hipbone and the temporal bone, respectively. Is there any connection between the shapes of the latter two bones?

Plate 51 is a picture of a left temporal bone. On plate 52 we see a skull indicating the position of the temporal bone. Perhaps we will get a first impression of the aforementioned connection on plates 53, 54 and 55. In the hipbone we recognize the joint socket of the thighbone and behind it the opening formed by parts of the pubic bone and the ischium. In the temporal bone we see, behind the joint socket of the mandible, the acoustic opening. The horizontal line of the back part of the zygomatic arch reminds one of the horizontal line in the hipbone. There is no other very obvious resemblance.

We can discern a stronger resemblance when we place the mandible beside the temporal bone and the thighbone beside the hipbone (pls. 56 and 57).

Shoulder blade and temporal bone

The question now is: To what is the upper jaw attached as seen from the side? The answer is: To the sphenoid bone. The sphenoid bone is hardly visible from the outside of the skull. Situated between the temporal bone and the

Pl. 51. Left temporal bone.

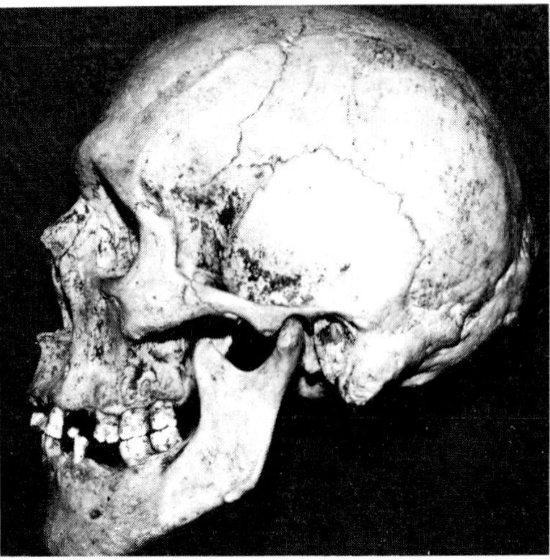

Pl. 52. Skull, indicating position of temporal bone, just above rear jawbone.

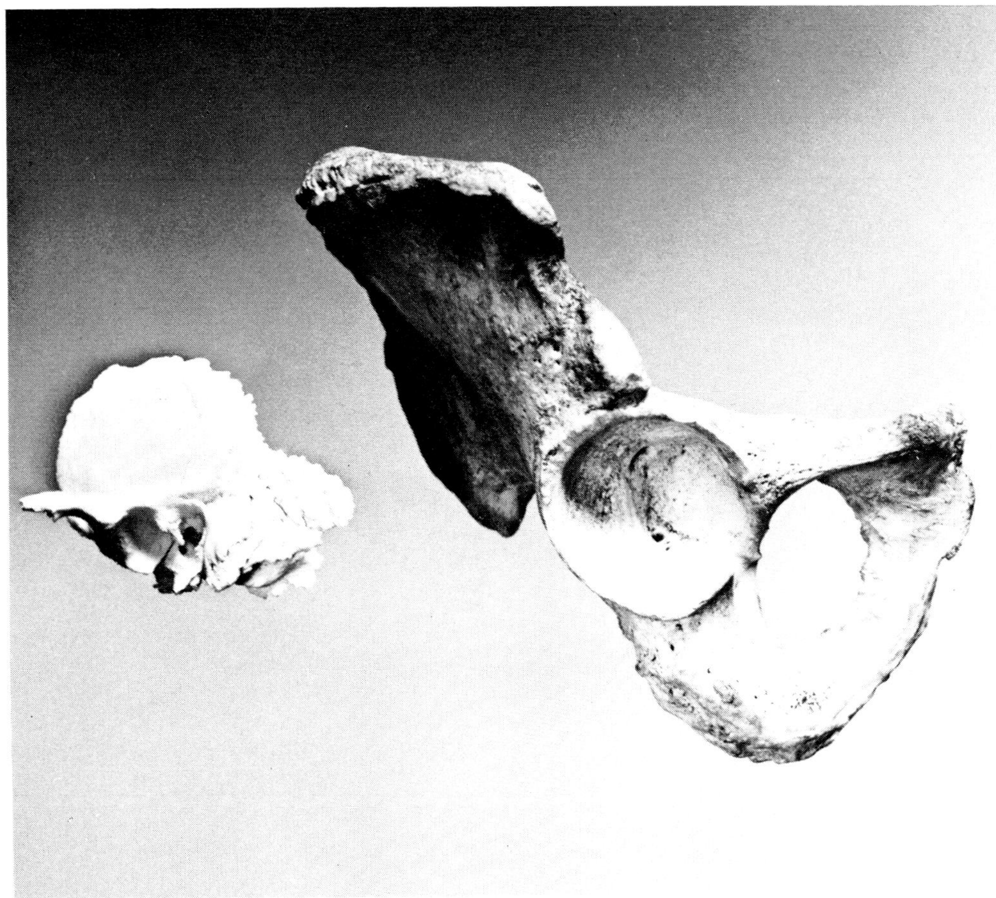

Pl. 53. Left temporal bone and hipbone.

frontal bone, it forms an important part of the cranium, to which we will return later.

To continue the comparison of the arms with the upper jaw, as we compared the legs with the mandible, we must keep in mind that the bone to which the arm is attached (the shoulder blade) will in turn be compared to the temporal bone. We just mentioned that the upper jaw is attached to the temporal bone by an intermediate bone (the sphenoid), but this is not detrimental to our further discoveries.

Let us now make a comparison between shoulder blade and temporal bone in a special position, shown in plates 58, 59, and 60. The resemblance is especially striking if one compares the edge where the two sides of the shoulder blade meet with the corner-ridge at the edge of the flat part of the tempo-

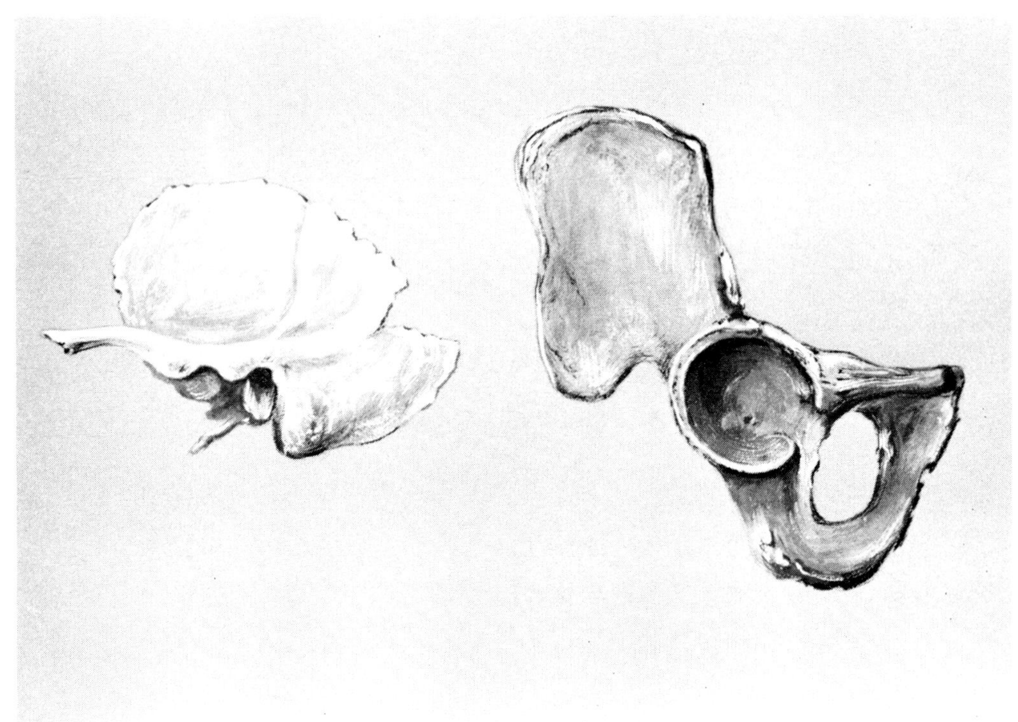

Pl. 54. Drawing of left temporal bone and hipbone.

Pl. 55. Drawing of left temporal bone and hipbone.

Pl. 56. Mandible and temporal bone.

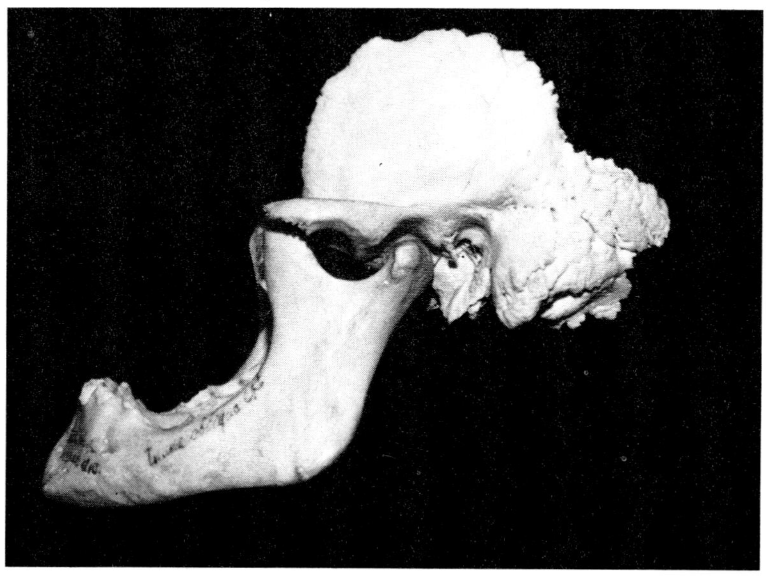

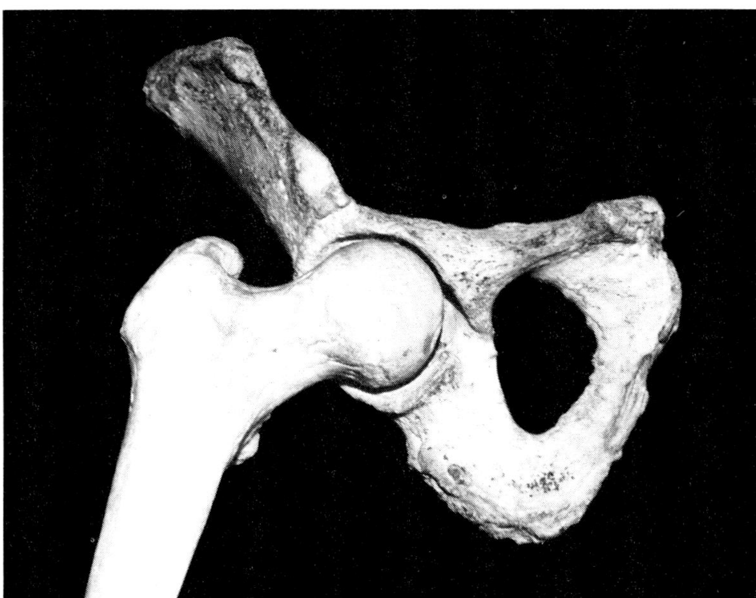

Pl. 57. Upper thighbone and hipbone.

Pl. 58. Temporal bone and shoulder blade.

ral bone, and also the transverse processes: in the shoulder blade the coracoid process and the end of the comb, the acromion; in the temporal bone the petrous bone, which contains the middle ear and the back of the zygomatic arch.

Plate 61 shows a lateral view of both bones: the shoulder blade upside down, with the striking horizontal line of comb, shoulder blade, and zygomatic arch. The illustrations are important. Later they will be the links between apparently random comparisons and what will assemble all these impressions into a constructive idea.

Shoulder blade plus hipbone, and temporal bone

By placing the shoulder blade upside down on the hipbone, the position in which we have compared it with the temporal bone, we can again expect a clearer connection between the two images; for we join together what each had, on its own, been seen as comparable to the temporal bone (pls. 62 and 63). The dorsal side of the zygomatic arch is now clearer. The shoulder blade

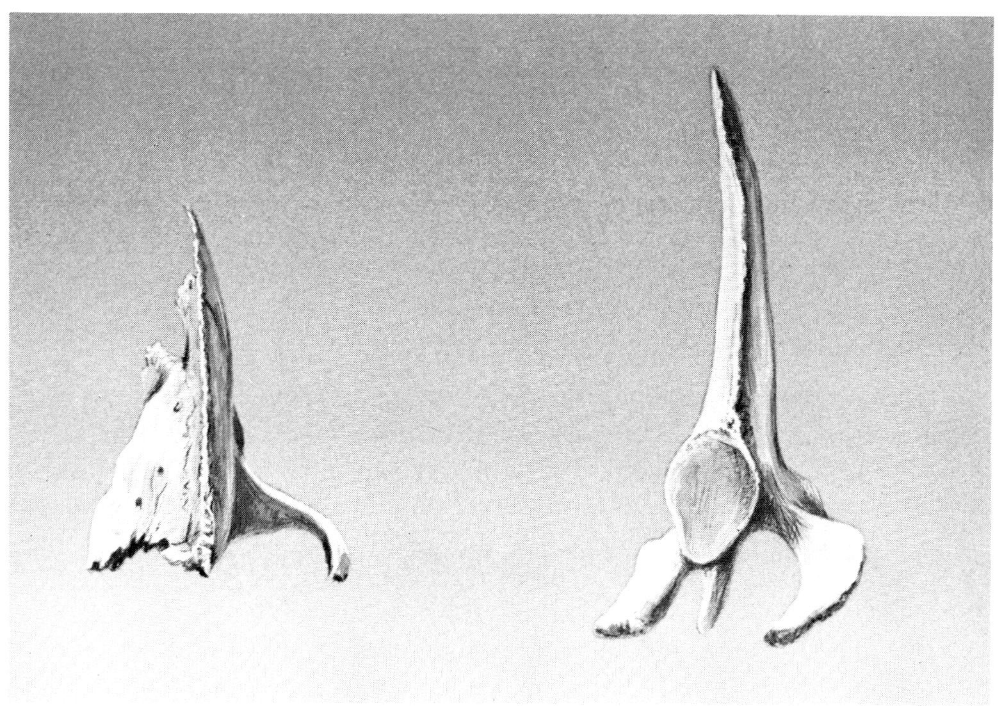

Pl. 59. Drawing of temporal bone and shoulder blade.

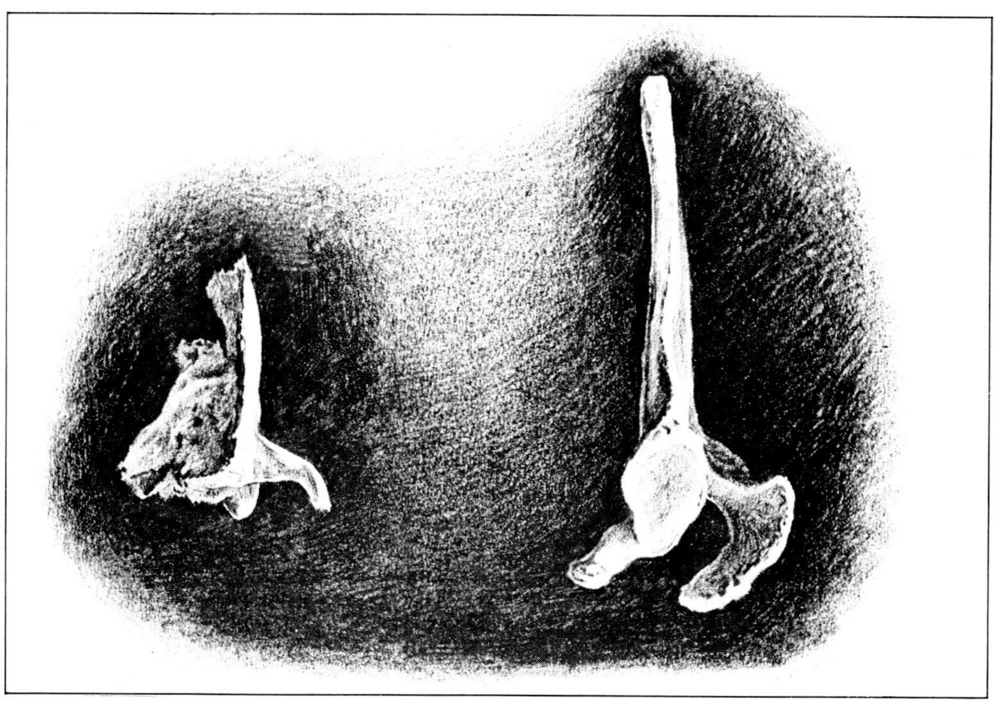

Pl. 60. Drawing of temporal bone and shoulder blade.

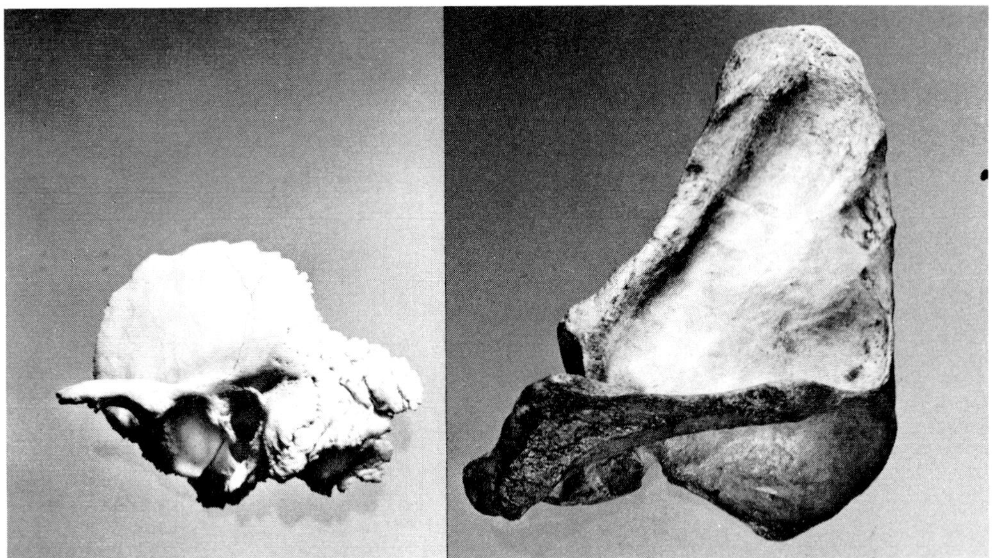

Pl. 61. Lateral view of temporal bone and shoulder blade.

is now more related to the leaflike upper part of the temporal bone; the hipbone is more related to the lower parts.

Shoulder blade plus hipbone plus thighbone, and temporal bone plus lower jaw

The picture becomes even more complete when we put the thighbone and the lower jaw in their allotted places. Three elements of the trunk skeleton will then be placed so that the comparison with the shapes of the temporal bone combined with the lower jaw will have a firmer foundation (pls. 64, 65, 66, and 67).

First we will reiterate what has been said about this part of the skull: the thighbone has been compared to part of the lower jaw (the whole leg with the whole of the jaw); therefore the hipbone, with which the thighbone articulates, must be found again in at least part of the temporal bone.

Finally, the upper jaw must also be considered as a limb, just like the lower jaw; in other words, we must try to see it as a transformation of the arm.

Because the arm articulates with the shoulder blade, we must also compare the temporal bone with the latter. Hipbone plus shoulder blade therefore equal the temporal bone; legs plus arms equal lower and upper jaw.

Thus we have found again the original theme of vertebra with rib, after

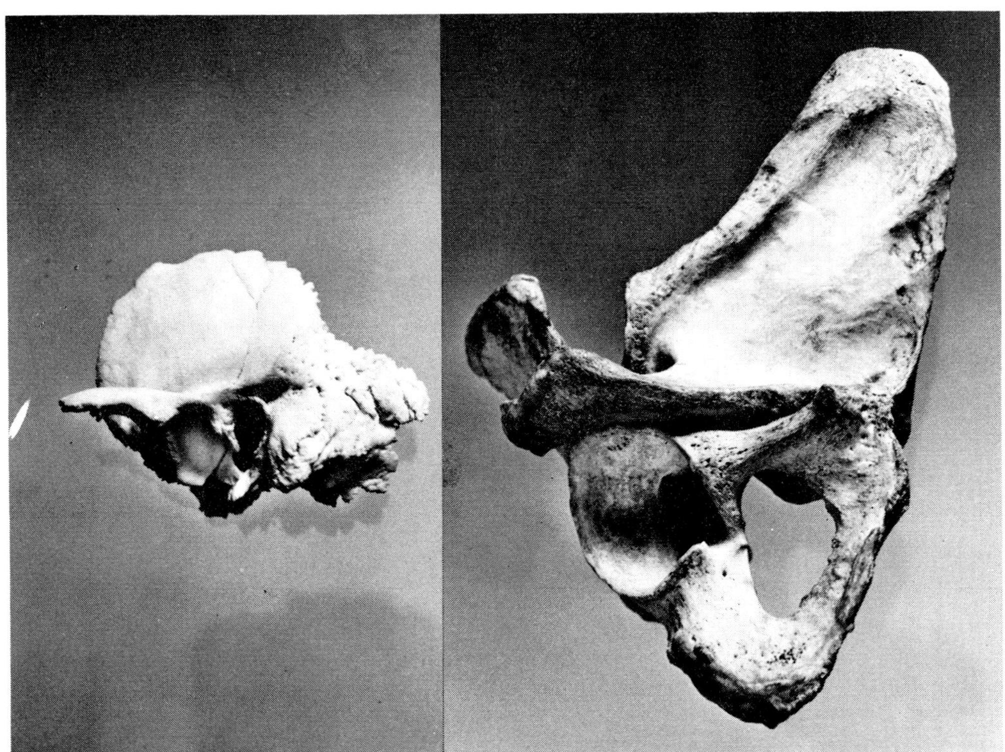

Pl. 62. Temporal bone and upside down shoulder blade above hipbone.

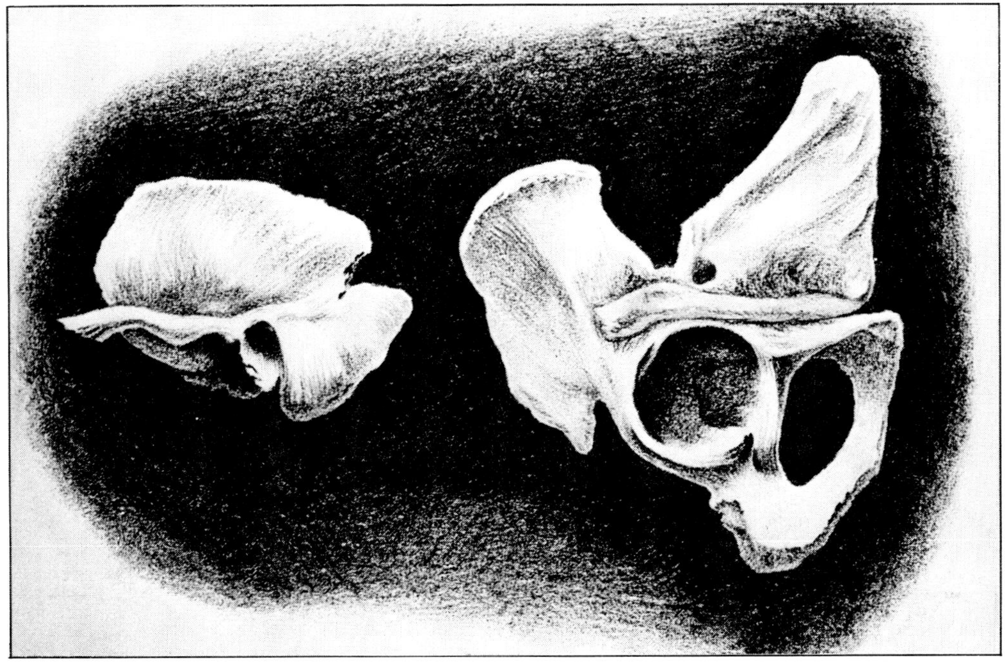

Pl. 63. Drawing of temporal bone and upside down shoulder blade above hipbone.

the metamorphosis in the shape of girdle with limbs, now for the third time in the combination temporal bone with jaws.

Pelvis and base of the skull

It is important that we understand our progress thus far. Up to now, we have spoken of a shoulder blade. However, we must stress that we have been dealing with the *right* shoulder blade and that previously we discussed the right hipbone and thighbone. In contrast, we dealt with a *left* temporal bone and the *left* half of the lower jaw. The shoulder blade, moreover, was placed upside down on the hipbone, in order to discover a resemblance.

Later, we will see how this unusual position of the shoulder blade can be integrated into our discussion. First we must attempt to justify the strange discovery that a group of bones from the right half of the body can be compared to elements from the left side of the head.

We must not forget that the same is true for the other side of the head. There we have to imagine a right temporal bone with a right lower jawbone and, next to it, the combination of a left hipbone, left shoulder blade, and left thighbone.

Now comes the question: What do we find *between* the two hipbones, and what do we find between the two temporal bones? In order to logically develop this train of thought, we must picture, next to the pelvis, the base of the skull, which can be seen by removing the dome of the skull. Pictures of the base of the skull in plates 71 and 72 show us what is meant.

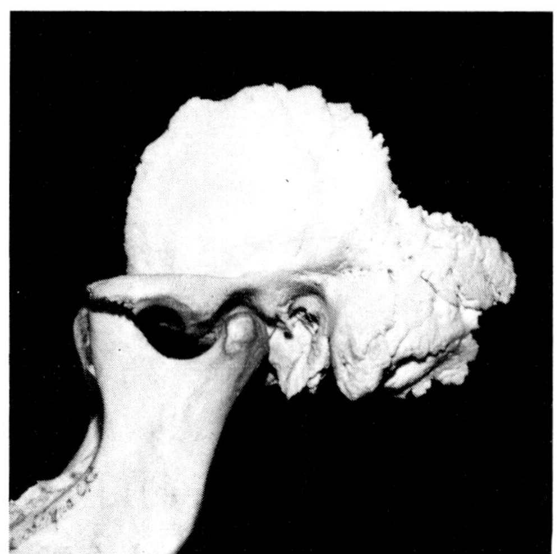

Pl. 64. Temporal bone with lower jawbone.

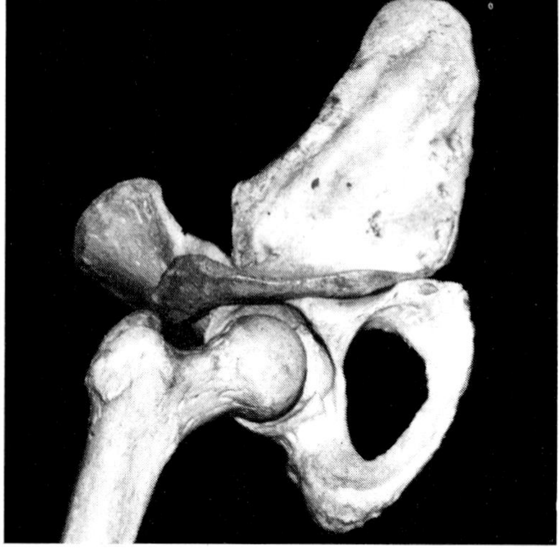

Pl. 65. Upside down shoulder blade with hipbone and upper thigh bone.

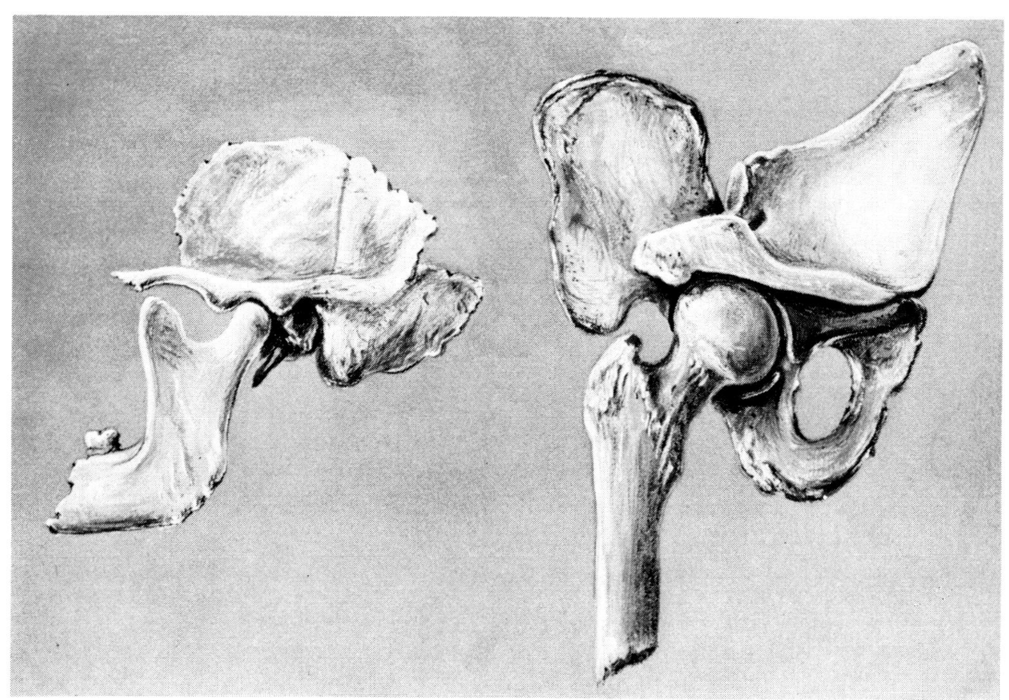

Pl. 66. Drawing of temporal bone with lower jawbone (left) and upside down shoulder blade with hipbone and upper thigh bone (right).

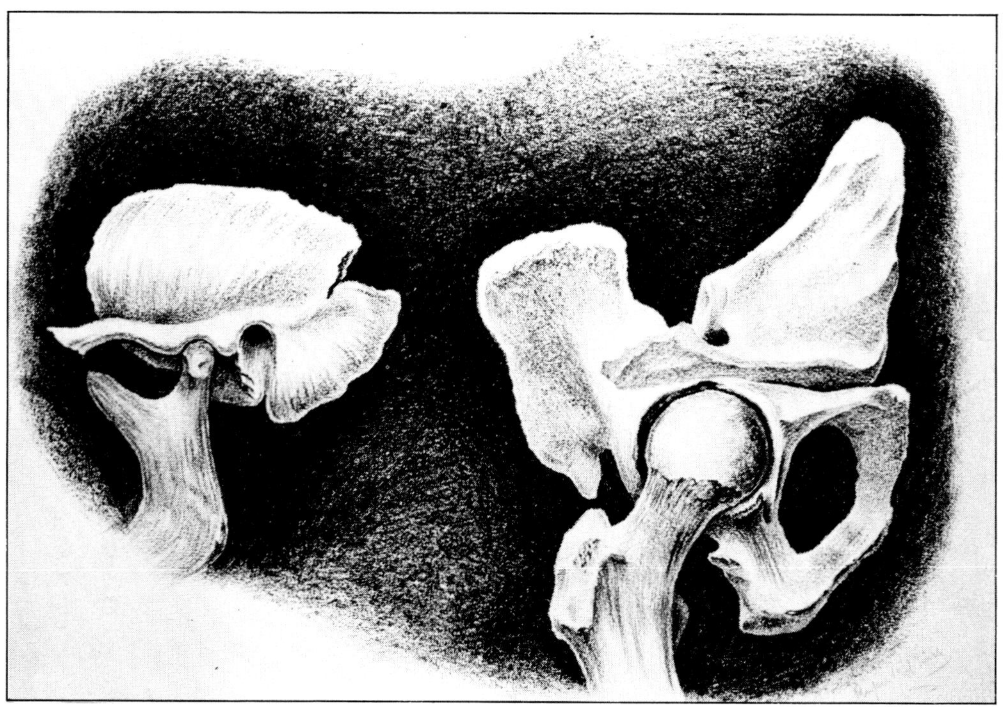

Pl. 67. Same as Pl. 66.

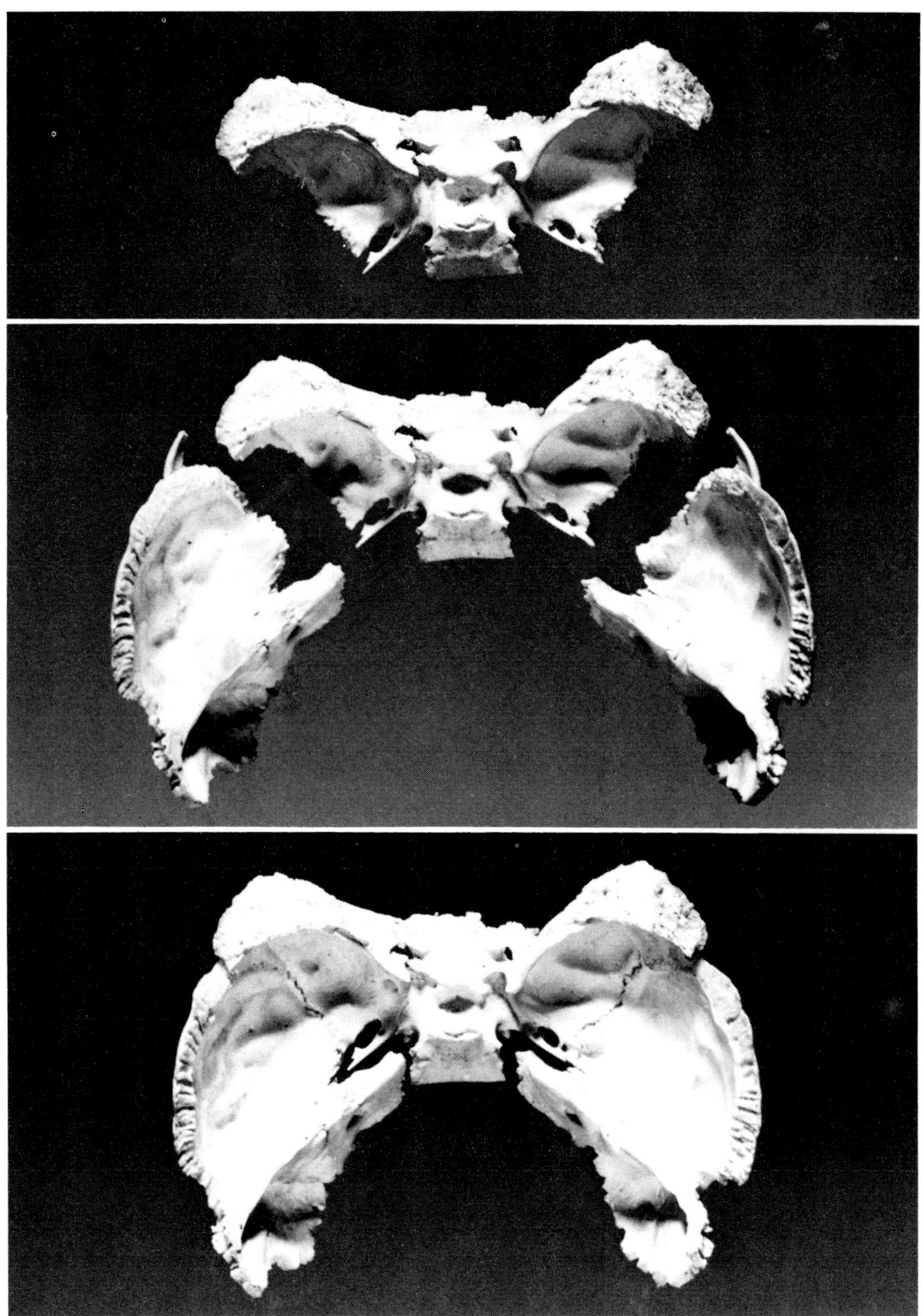

Pls. 68, 69, & 70. Right and left temporal bones and sphenoid bone.

Here we see the sphenoid bone between the temporal bones. For the sake of clarity it has been pictured separately, next to the base of the skull, once more below with the temporal bones next to it, and finally, connected with the temporal bones (pls. 68, 69 and 70).

When we think of the hipbones, which we have compared with the temporal bones seen from the side, and again ask ourselves what is to be found between the hipbones (that is to say, what encloses the pelvic cavity), we find at the back the sacrum—the part of the spinal column that consists of five fused vertebrae and is part of the pelvic ring (plates 73 and 74).

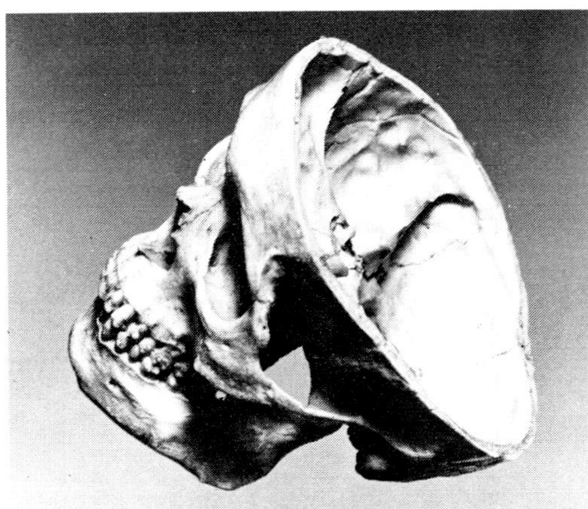

Pl. 71. Lower skull with dome removed.

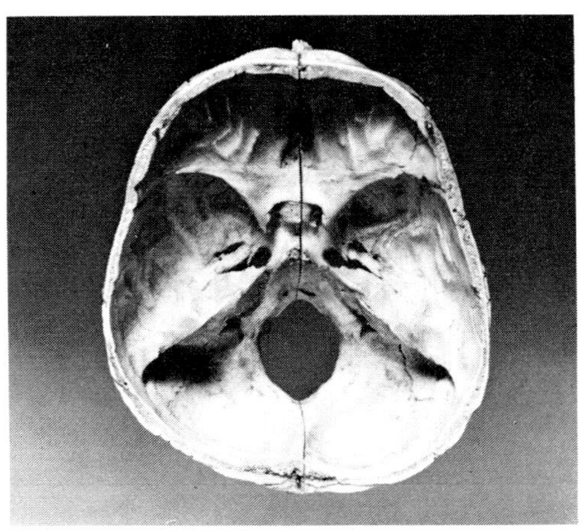

Pl. 72. Lower skull with dome removed, viewed from above.

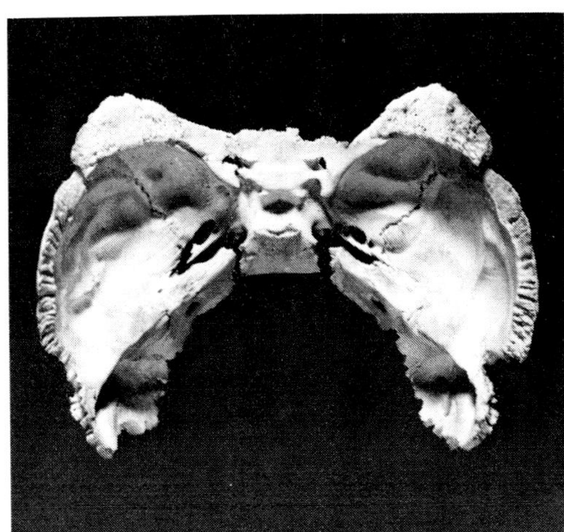

Pl. 73. Right and left temporal bones and sphenoid bone.

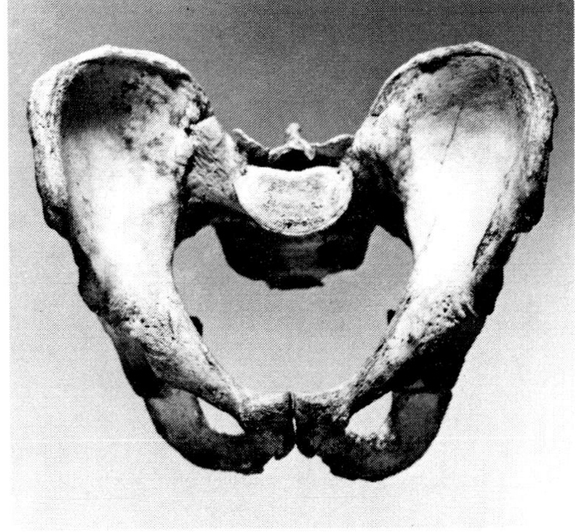

Pl. 74. Hipbones and sacrum

Seen as a whole and visualized in a position that corresponds to the comparison of hipbone with temporal bone, the pelvis faces backwards. The front of the pelvis points in the direction of the occiput; thus in the picture, it is placed to the right. The pictures of the temporal bone with the lower jaw, however, show parts of the skull that point to the left.

First we look into the base of the skull, then we look into the pelvis. Thus one can compare plates 73 and 74. It is striking that this recurring resemblance links directly with our previous question: Is there a coherence in the shapes that are found between the bones and the bone combinations we have just studied?

In plate 76 the pelvis is shown so that we can see the sacrum at the back, between the hipbones. We see a number of openings through which nerve bundles run from the spinal cord to the pelvic cavity.

Pl. 75. Drawing of right and left temporal bones with sphenoid bone (left) and hipbones with sacrum (right).

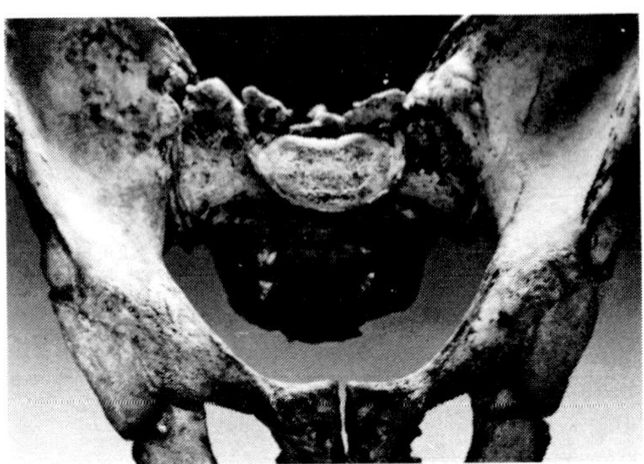

Pl. 76. Pelvis, showing hipbones and sacrum.

Plate 75 has been drawn in such a way that the similarity with a number of openings in the sphenoid bone becomes clearer and that the comparison with the dorsal wall of the pelvic girdle appeals to us more directly.

Clavicle and cheekbone

One will have noticed that there has been no mention so far of one part of the shoulder girdle; namely, the clavicle.

Going back to the positioning of the shoulder blade in plates 65, 66, and 67 (the three elements together), the clavicle should be seen as a prolongation of the horizontal line that in the temporal bone forms part of the zygomatic arch. This zygomatic arch continues in the cheekbone. This can be checked by placing one's finger on the lower rim of the eye socket, formed by the cheekbone, and tracing it backwards. According to our observations, cheekbone and clavicle ought logically to be seen as each other's equivalent. The cheekbone has, nevertheless, a totally different shape; this is connected, in the first place, with the totally different orientation caused by the nose and eye areas, as is the case with the upper jaw. All the same, it is possible for the attentive observer to discover a certain affinity of shape. (See plate 52, in which the zygomatic arch and the cheekbone are clearly discernible. Plate 77 shows two clavicles).

Pl. 77. Two clavicles.

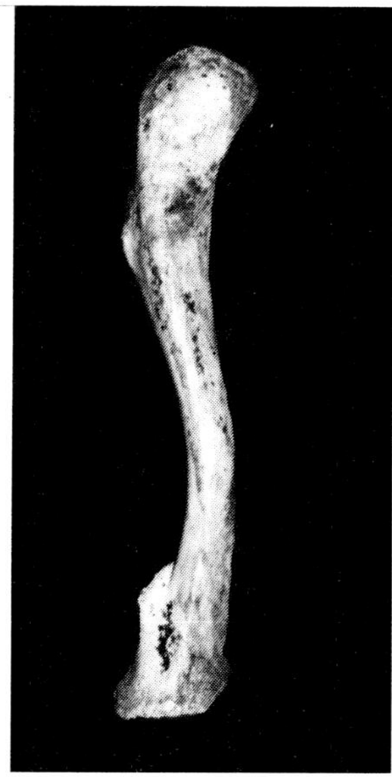

Pl. 78. Clavicle.

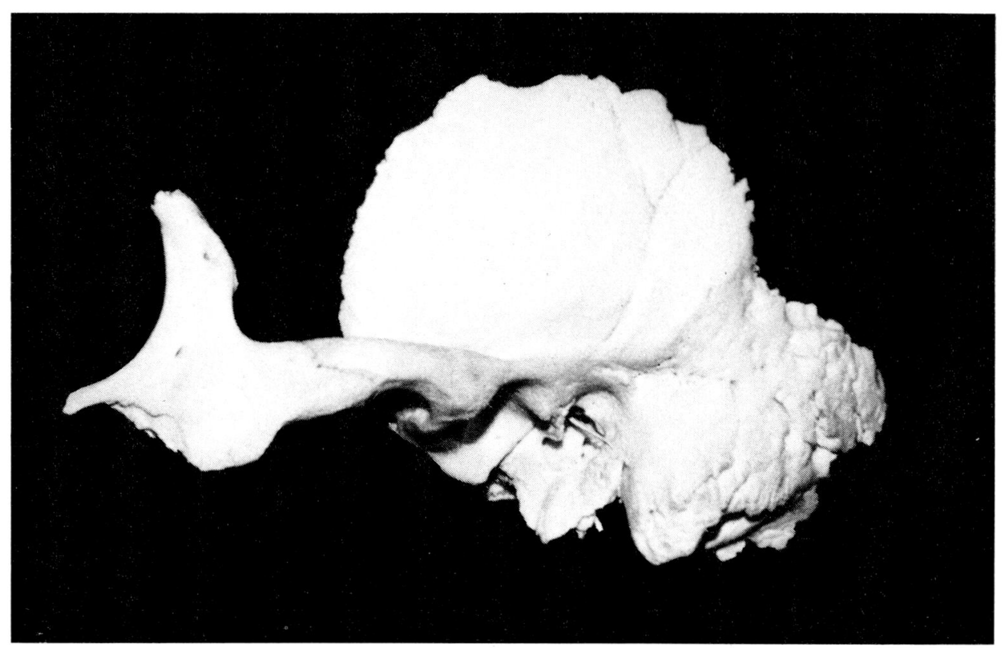

Pl. 79. Cheekbone and temporal bone.

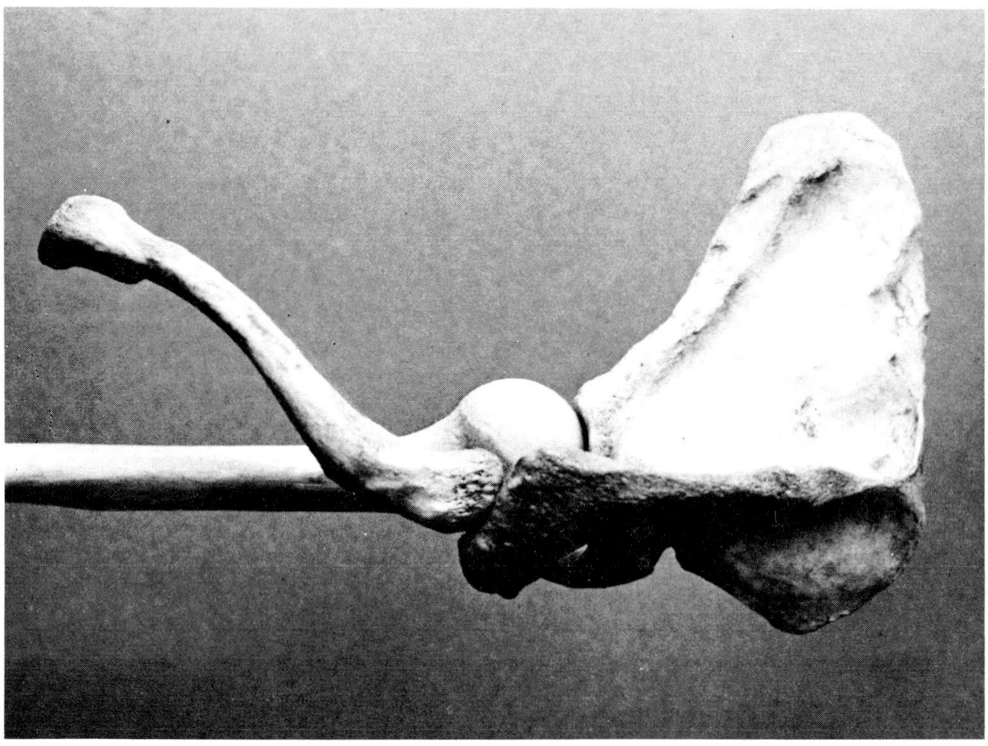

Pl. 80. Clavicle, humerus, and shoulder blade (scapula).

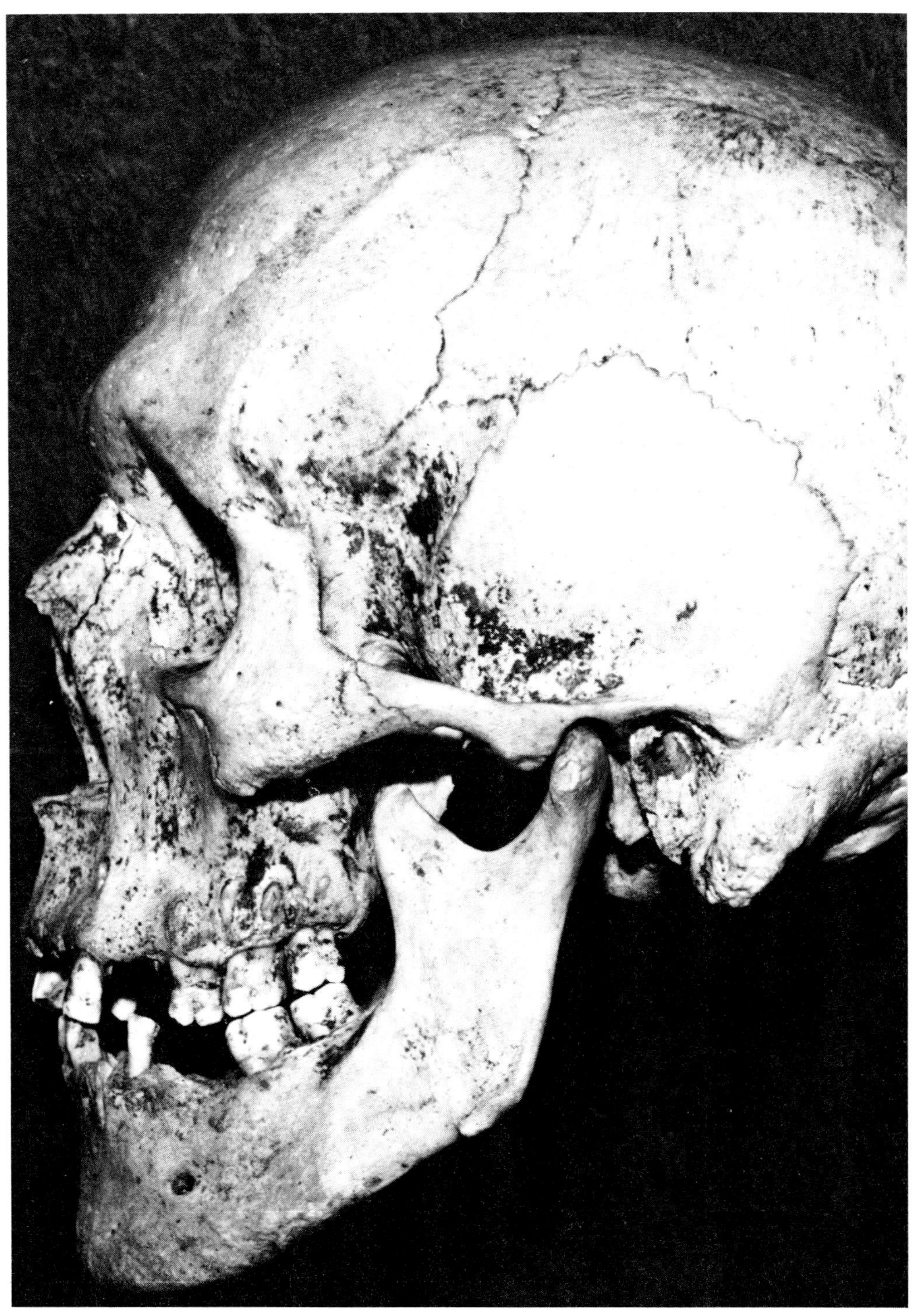

Pl. 81. Complete skull, left side, showing cheekbone, temporal bone, and jawbone.

Plates 79 and 80 show us the connection between clavicle and cheekbone, clearly depicted in plate 81. For the sake of completeness a humerus has been attached to the shoulder blade in plate 79. This represents the principle of the upper jaw in plates 40 and 50. Plate 78 shows the clavicle in a vertical position. Here it is suggested that one might be able to discern the whole human form even in the clavicle—which is, after all, a long bone.

It will have been noticed that not all the skull bones are represented in our discussion. Indeed, the examples already presented are intended merely as a first tentative scanning of the mystery of the human form.

Chapter 4
Viewing our discovered phenomena in a new light

Surprising information

What conclusion can be drawn from all this? In the first place, we can say that metamorphosed shapes from the axial skeleton can be recognized in the shapes of the skull bones. Just as the girdles and the limbs were a metamorphosis of the vertebrae-ribs units, so we can conclude that an archetypal theme—a vertebra with two ribs—transforms itself into the girdles with the limbs and finally into the central part of the skull, the temporal bone with the jaws.

We must increasingly ask ourselves where all this leads. Up to now, we have talked about more or less striking similarities of shape. We have shown that the axial elements and skull elements between which comparisons were made were viewed from opposite sides.

All this forms a meaningful whole only if we concern ourselves with what Rudolf Steiner has to say about the human form after death. This is not the place to discuss in detail the emotions one can feel when hearing this; for many people, there is a void before birth and after death. A few points, however, must be discussed.

The first question is: How did Steiner arrive at ideas that seem impossible to know—impossible, that is, with only the aid of our sensory perception. However, there is another path of knowledge, and Steiner has mapped it out.

Steiner's incredibly voluminous oeuvre (more than 300 books) is available to all. Everyone can read his work and get some idea of what it is all about.* Nevertheless, we will give a brief answer here to our question.

*Some of the most important books by Rudolf Steiner:
How to Know Higher Worlds; Intuitive Thinking as a Spiritual Path; Theosophy; An Outline of Esoteric Science (all published by Anthroposophic Press)

We live in a world of phenomena that present themselves to us as observations. We are inclined to say that something does not exist if it cannot be observed. We encounter this very clearly in the enigma of the existence of the human being before birth and after death. People say that this is a matter of faith. Thus birth and death—and also waking and falling asleep—come to stand outside the limits of human knowledge. Anything that exceeds these limits is called fiction. Faith is fiction.

In general, people do not fully realize what this means. If someone tells me that he has two dollars in his wallet and I say "I believe you," it means: I assume you speak the truth. But one would commit a grave injustice if one viewed the accepted notion of faith, as it exists in the form of many different religions, as an assumption in this same sense. It would be peculiar, to say the least, that people have been burned at the stake and thrown to the lions for the sake of such a casual assumption.

Until recently faith was an experience shared by many. It was indeed in the first place called a matter of sentiment. Only in the last few decades has the emphasis shifted toward impressions of sense perception, toward learning to distinguish. That extraordinary feeling of confidence inherent in faith gradually began to disappear when the new attitude grew more popular. In fairness, one must admit that further back in time the experience of the reality of faith must have been much stronger. Old Testament stories, such as Jacob wrestling with an angel and Samuel hearing his name called in the night, can be seen as illustrations of this deeper experience, even if one does not accept the descriptions literally.

It would be wrong to say that we have exchanged a former vagueness for the new certainty of our present consciousness. The enormous number of modern facts hides the uncertainty on which these facts are based, and that each can discover for himself. The feeling of confidence linked with the old faith has been replaced by a hypothesis; in other words, theory has replaced theogony. Instead of the word creation, we have the word fact (by derivation only another word for creation); instead of faith, we have the demand to see before believing.

A question living in many people today is whether it is possible to find a way back to that idea and experience of a reality that, until a short time ago, could be achieved by faith, while at the same time preserving the gains made by science and the analytic mind. This question is answered by anthroposophy.

Rudolf Steiner speaks about observing realms beyond the sense perceptible world and therefore outside the known limits with the same certainty that we talk about our surroundings. He stresses the fact that our ordinary senses can naturally provide us with only that image of the world known in everyday life. It is necessary to develop other organs of observation for that other, spiritual world. This, however, can only be done in our familiar world of observation, which is after all nothing but the result of the activity of the creative world itself. The exercises that can bring this about and the conditions that have to be fulfilled are discussed at length in anthroposophy.

The results of the study of the spiritual world, which might also be called the world of the creative beings, have been described in several works by Rudolf Steiner. The obvious question is what to do with communications that we cannot verify, at least not at present. Rudolf Steiner answers this question clearly in his book *Knowledge of the Higher Worlds and its Attainment*:

> In order to establish the facts through research, the ability to enter the supersensible worlds is indispensable; but once they have been discovered and communicated, even one who does not perceive them himself can be adequately convinced of their truth. A large proportion of them can be tested offhand, simply by applying ordinary common sense in a genuinely unprejudiced way. Only, one must not let this open-mindedness become confused by any of the pre-conceived ideas so common in human life. Someone can easily believe, for example, that some statement or other contradicts certain facts established by modern science. In reality, there is no such thing as a scientific fact that contradicts spiritual science.*

This means that we may never believe in advance a statement emanating from this philosophy but at the same time we may never *not* believe it. We can listen without bias to a statement and see if it answers a given question.

To arrive at a clearer understanding of what is at stake here, it would be wise to seek a connection between such statements and our question: What is the law behind the prodigious discovery of the metamorphosis of the axial skeleton into the skull?

The pertinent statement is that the shapes of the head are a metamorphosis of the shapes of the axial skeleton of the previous incarnation. In other words, our trunk becomes our head in our next life. This means the head we had in this life disappears after death. Therefore, the trunk we are born with is new. This has nothing to do with the mortal remains of the body.

*Rudolf Steiner, *How to Know Higher Worlds* (Anthroposophic Press, 1994), formerly entitled *Knowledge of the Higher Worlds and Its Attainment*

The enigma of our body's shape

From birth to death the substance of our body is held together by forces. A good illustration is to think of the moment of death, the actual dying, when those forces withdraw and matter obeys the laws of so-called dead nature (the mineral realm) to which it reverts. The body is no longer held together and one speaks rightly of decomposition.

Medical science offers us another way to come closer to understanding the sum total of those forces. To the question concerning the origins of illness in general, the answer is usually clear: chemical, physical, biological, etc. The question concerning the healing of a wound, for example, is often answered by pointing out the wonderful *vis medicatrix naturae*, the natural healing force. This answer either shifts the problem and gives no answer at all; or it gives the answer we mean: the sum total of the forces that maintain our form during our lifetime.

We are dealing, therefore, with forces that shape our body and create its structure. Rudolf Steiner called the total of these forces the "body of formative forces." All living beings—plants, animals and humans—have such a body of formative forces, in earlier times called etheric body or life body. The deep rift between what the human being experiences as soul and spirit and what is described as body in anatomy and physiology has been caused by the fact that this life body is generally not considered an element of the human constitution nor studied as such.

When we have familiarized ourselves a little with such a thought we can develop an understanding for what has shaped the skeleton and the whole human form in the course of the embryonic development and remains active in those forms all through life. At the moment of death these formative forces withdraw into that world from which they acted.

What we have learned of the visible forms must become a guide to our better understanding of what takes place in the spiritual world after death.

Our bodily shape after death

Just as the shape of our body and skeleton is the result of the activity of the body of formative forces, so it is understandable that the origin of the human shape must also be found in that realm out of which these forces work. For those who are able to observe in this realm, our bodily form does not disappear immediately after death. Although freeing itself very quickly from the spatial element, it can still be followed for awhile. At that stage the head fades away and disappears; the head does not go any further. The rest of the form expands more and more at first and is then slowly regathered by those beings who created it.

When a new incarnation is being prepared the developing form appears gradually in the same way as the previous one had disappeared—another perceptible body of formative forces comes into existence. There the trunk-limbs pattern of the previous incarnation becomes recognizable in the nascent shape of the head that slowly affirms itself. Therefore, the trunk we are born with later must be considered as a new addition. It comes into being in the embryonic development as the result of cosmic creative powers and the hereditary force active in the mother. The human ovum developing in the mother's womb brings with it its own essence, in contrast to the plant and the animal where each later form is completely defined by the characteristics of its species. There is, of course, an interaction between that individual element of the human and the inherited characteristics of the mother.

Thus heredity alone does not define the human being. This alone makes it possible for new human elements to develop that link up with previous ones.

This discussion will be of value only for those who are prepared to at least listen to such thoughts and to see what new ideas could develop in them, if they are willing to let their thoughts be fructified by communications from spiritual science, as Rudolf Steiner sometimes called anthroposophy.

Trunk and skull

How can we visualize this metamorphosis of the trunk into the head of the next incarnation?

Let us return once more to the human form as a whole as it appears in the skeleton and imagine that by bending it forward we roll up the skeleton in such a way that the shoulder blade rests on the upper edge of the pelvis. The head must, of course, be thought of as absent. Ten Houten de Lange, who took a lively interest in this subject, has taken pains to make schematic drawings (plates 82, 83, 84 and 85).

Special attention should be paid to the shoulder blade that, if one visualizes this movement, does not make a single turn but ends up standing on the pelvis, just as is shown in the drawing. In the last drawing, which represents the final phase, the position of arms and legs is rendered in such a way that the comparison with upper and lower jaw in the skull becomes most apparent.

For the sake of clarity plate 86 represents the skull viewed from the left next to the drawing of the rolled-up axial skeleton, while in plate 88 the corresponding shapes of the skeleton on the skull have been drawn in projection.

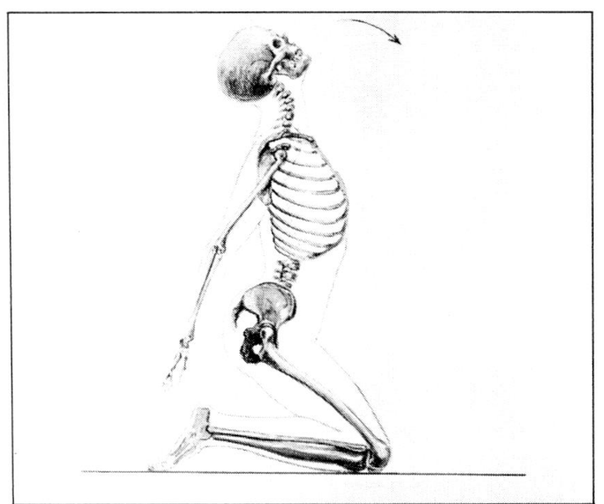

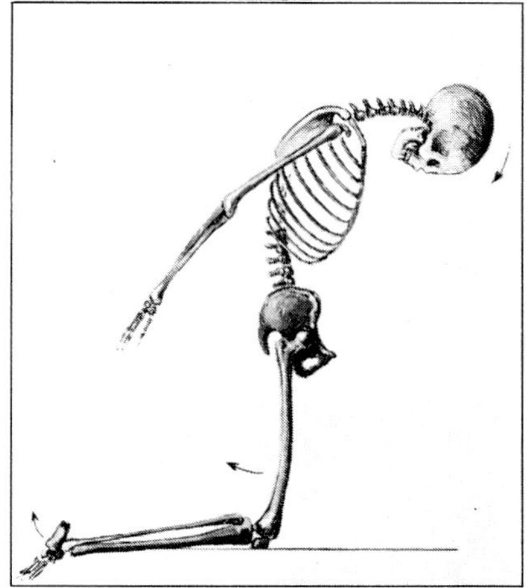

Pls. 82 & 83. Skeleton rolling forward.

Plate 87 repeats plate 85. We are reminded that a remnant of the limb principle, namely, the fingers, can be found in the grooves of the jaws where the teeth are situated.

The hollows of the hand and foot together thus become the hollow of the mouth. Plates 89 and 90 try to show this more clearly. While it has been said previously that the milk teeth can be seen as a metamorphosis of the ten fingernails and the ten toenails, we are now thinking of a metamorphosis that takes place in two successive lives, as we have encountered it in the trunk and skull.

When one again considers the rolled-up trunk and examines exactly how the hands and feet should be placed against each other to turn them into "jaws," one can understand why hands must have the little fingers and feet must have the big toes.

I don't think the fact that we have far more teeth after shedding our milk teeth need bother us. After all, we are dealing here with the discovery of hidden laws that are veiled by changes having a totally different origin than the one that has preoccupied us until now.

The permanent teeth do not complement the milk teeth. Milk teeth are completely rejected and thus represent a world on their own, a world of the past that has its own language. It might be suggested that the permanent teeth should be viewed as an activity of the life of the new trunk; that is, of the new body. Here the laws of body and head meet each other.

In addition to what we have just said about the hollows of hand and foot being the hollow of the mouth, a further conclusion is yet to be drawn.

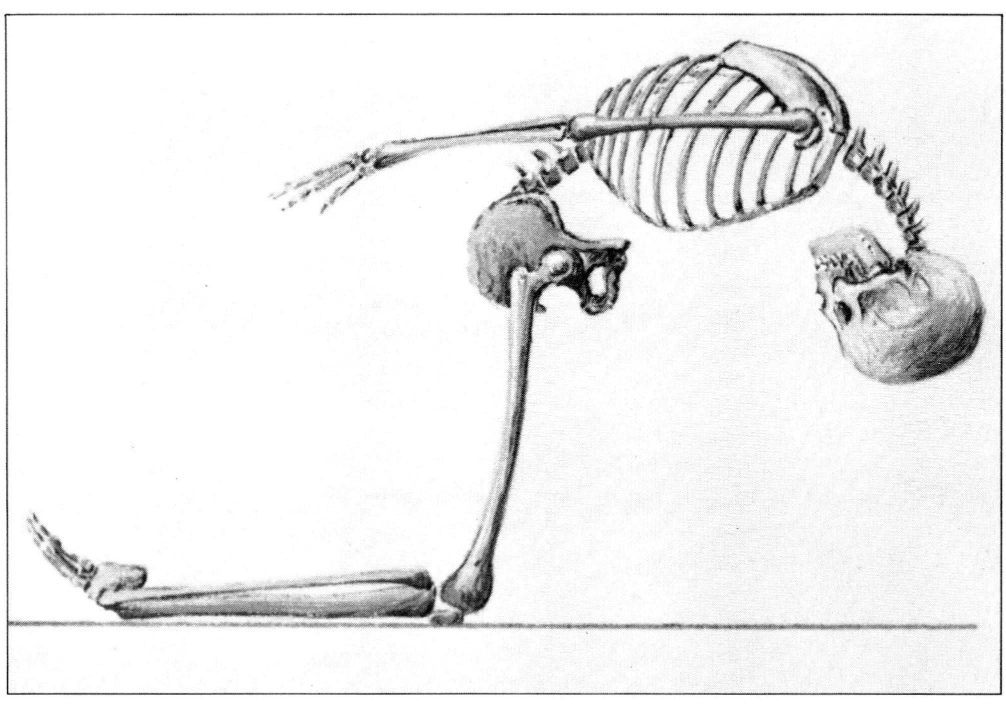

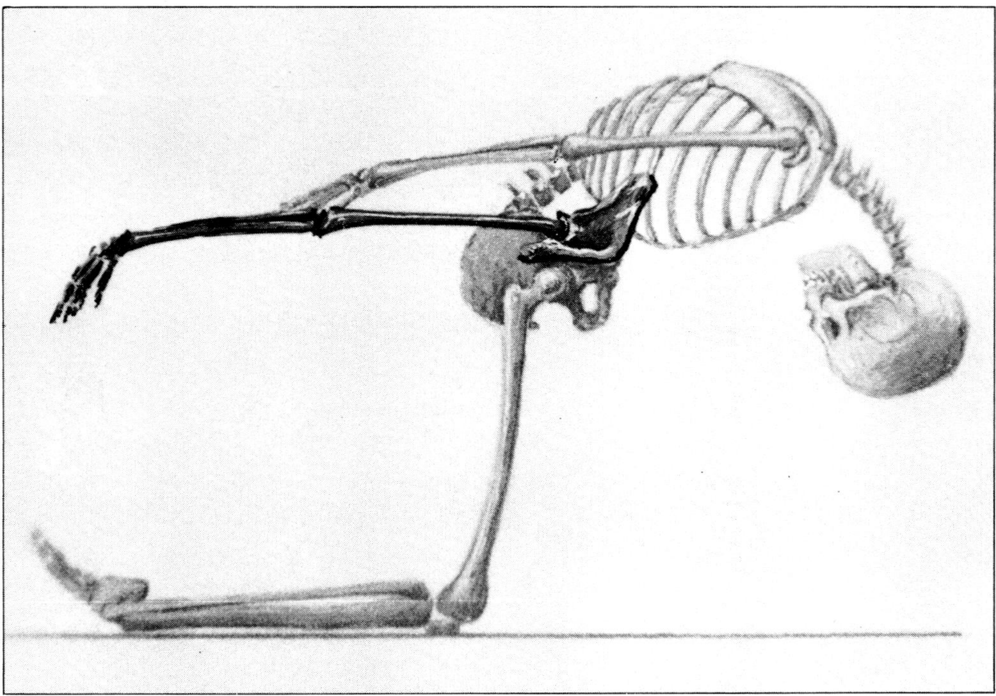

Pls. 84 & 85. Skeleton rolled forward with shoulder blade on upper edge of pelvis.

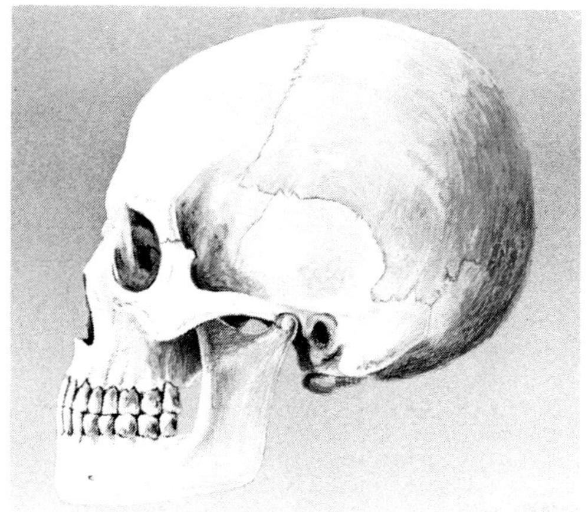

Pl. 86. Skull seen from the left.

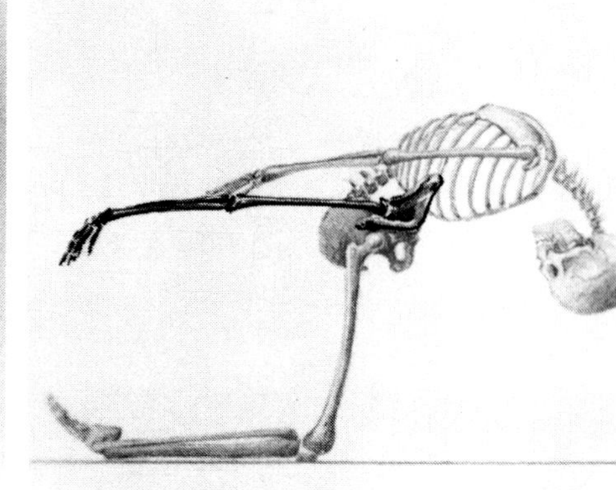

P. 87. Skeleton rolled forward with shoulder blade on upper edge of pelvis.

In this book we have talked mainly about the shapes of the skeleton. The idea "trunk becomes head" is not valid only for the skeleton. The forms and "characters" of other members of our body, of our organs, of soft parts, and so on, also take part in this process.

Trying to find an equivalent in the head for each organ or place in the trunk brings the risk of losing oneself in speculation. By becoming more and more acquainted with this train of thought, one develops a feeling for what is justified and what is not. Before one arrives at a real understanding one can only experience impressions and suppositions.

Thus I cannot escape the idea that the comparison of hand-plus-foot-hollow with mouth-hollow contains a deep truth. Could there be a connection between the deeds hands and feet have accomplished in one life and the words spoken by the mouth in a subsequent life?

In one of Grimm's fairy tales, "The Three Goblins in the Wood," there is an episode that can be viewed in a special light. In the tale are two girls with entirely different characters. One is lovely and sweet, the other ugly and nasty. The nice girl is sent into a large wood in the middle of winter while it is snowing. She goes into a little house inhabited by three goblins. She is friendly, polite, and generous. When she leaves she receives a gift—for the rest of her life a gold coin will fall from her mouth whenever she utters a word. The nasty sister also goes into the wood, visits the house of the goblins, and is rude, unfriendly, and mean. Her fate will be that with every word she speaks a toad will jump out of her mouth.

How are we to interpret this? People who have led a useful life are born

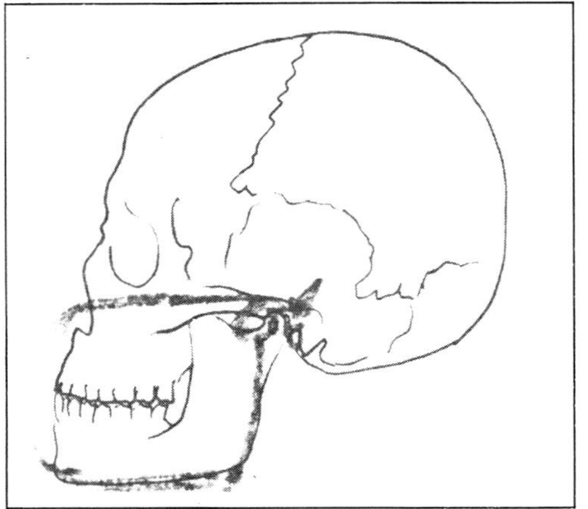

Pl. 88. Skull seen from the left with corresponding skeleton shapes drawn in projection.

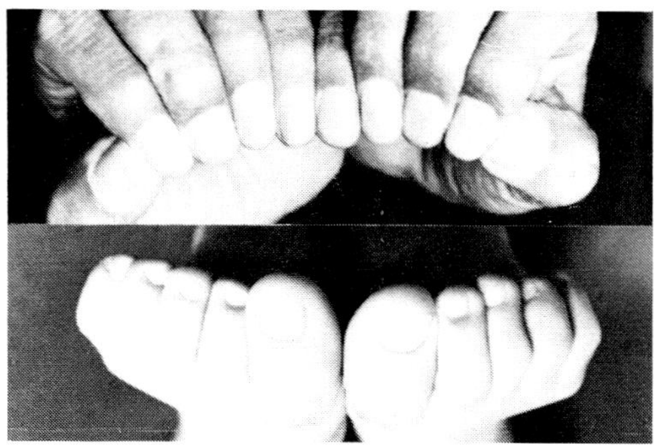

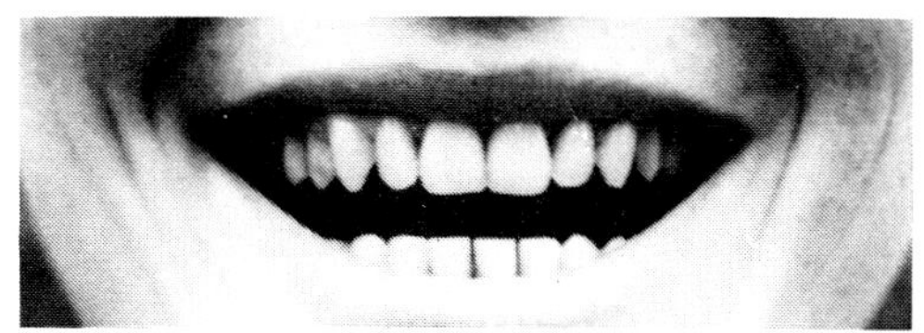

Pl. 89 & 90. Ten fingernails and ten toenails compared to milk teeth.

with a different way of speaking from those whose actions were dictated by purely selfish motives. How this develops in the course of life is not decided beforehand, naturally.

If the limbs become jaws and thus lose part of their mobility, especially the upper jaw, one can imagine a metamorphosis of the muscles of the limbs into the muscles of the face. What was a gesture now becomes mime! One can go even further. Our gestures are also, up to a point, a kind of talking. We speak of gesticulation, the language of gestures.

What purpose do our teeth serve? Biting, of course, springs first to mind. But let us not forget that our teeth play an important role in speaking. The consonants S and T, especially, are mutilated if we don't have our teeth.

Finally, is it not possible to think of the heart—as an organ between arms and legs—continuing to live as a metamorphosis in the tongue between the jaws? Both are composed only of muscles. Both organs never grow tired in our healthy lifetime.

Thus, when studying the body, one will come upon more and more impressions that are inclined to link with earlier findings. Those who, through pictures in books or while assisting at an operation, have seen the convolutions of our intestines will be reminded of the convolutions of the brain. In the first instance we encounter an area of warm, continuously moving vitality; in the second, an atmosphere of cool calm, coupled with a very low vitality. Our central nervous system is the area with least regenerative power. Here, too, we must concentrate solely on the impression through which one recognizes the law of metamorphosis in the human shape.

We have already said in the beginning that we are not dealing with facts, but with questions. This is why such a train of thought is perhaps less problematic than one might first assume. That which is doubtful for one person may be a source of deep joy to another.

Let us recapitulate: the trunk, rolled up and turned back to front, becomes the head of the next incarnation. This is the short cut to the result of our research.

Individuality and heredity

Again and again conceptions are brought forward that view the human being as only the most highly organized animal. It is good to point out that the line of thought developed here is not applicable to animals.

An animal is always a specimen of a species. One can study the life of an animal, for example that of a lion living today; yet one could have done so a hundred years ago. When the lion which is now alive dies, I will in a hundred years still be able to study its life. Thus in this respect the lion does not die. Of course this specimen dies, nevertheless "the" lion, "the" ant, "the" fish, the species remains in existence.

One might now propose that the human being, too, can always be studied afresh. Yet in our previous consideration, the crux of the matter was not the human being as species, but as individuality. In the chapter about the metamorphosis of animal and human being is a description of this.

Linked to this is the fact that the bodily form of an animal is completely determined by heredity. An animal is the consistent representative of its species. This is really the same thing that was claimed by saying the animal is a visible desire.

We have seen that the human being bears something within himself, by virtue of which he is radically different from the animal. Imagine the erect, speaking, thinking human being and next to that the horizontal, sound-making, instinct-gifted animal. One can then perhaps begin to do something with the concept of the human being as a thrice-raised being, three times redeemed from being tied down to the earth. Hence, his body becomes an instrument that can be used or played by an entirely different element—the human individuality.

One can thus develop a sense for the personality a person bears within himself in a particular life as being the revelation of something that shows itself only once in this form. A biography is only once! It is clear that what was just said about the animal does not apply here.

Nonetheless, we all know that the human body also obeys the laws of heredity. A direct expression of this are the forms and functions of the human body which all people have in common. This is the reason why it is necessary to discuss the interaction between the stream of heredity and the individual personality. Two laws penetrate each other in every incarnation. The individuality meets with a body that possesses hereditary traits.

What is individual in the human being expresses itself in the gifts he has for one thing or another, his talents. It is known that a close connection exists between particular talents and one's bodily structure. It is also known that someone who might be exceptionally well-suited physically for something does not necessarily have the corresponding talent.

It is different with animals. We have characterized the animal as a unity of desire, form, and environment. Here too one can speak of talent, as well as of instinct and desire. Aren't the incredible gifts of birds, fish, and kangaroos —to fly, to swim, and to jump—visible, completely expressed talents? We can arrive at a very unusual conclusion if we survey the entire animal kingdom and observe the great variety of talents displayed there. The conclusion is: Animals do not have talents, they *are* talents! The far-reaching one-sidedness of the animal talent, the recognizability of the animal in each part of its form, and the specific environment of the animal are united in one concept.

Now, how is that with we people? The human being *has* talents, yet this implies that he has something else—something said to be but a temporary guest on the earth. This is his individuality. The talents that this individuality brings along with it (how often it is said of a gifted person that he was given talent at his birth) and the personality could never have been inherited. What is indeed inherited are the corresponding physical qualities.

So these qualities enable a person to display a talent that he "brought with him." The conviction that these talents were inherited stems partly from the many talented people in one family. This, however, can be seen in a totally different light. As soon as we understand that the display of a talent needs a bodily structure that is inheritable, it is comprehensible that souls which are going to incarnate with that talent will be drawn to the parents who have this possibility. We can in special cases imagine a real waiting line in the spiritual world. The Bach family is a striking example.

If a person finds a hereditary structure that enables him to live out his talent and predisposition, it need not be that the parents, who have provided for this inherited form, show the same talent. It is of course possible that they do. The origin of such a structure may lie generations in the past. If, on the other hand, the body provided by the parents does not possess the structure needed by the talent in order to unfold, then it is possible that this predisposition either cannot develop itself, or develops only partially. It can happen that a hereditary structure is not suited for manifestation of the individuality at all. The form that one then finds is not for nothing called a deformation. This is the case with children that are born mentally handicapped. In reality, the spirit itself can never be "handicapped." Here, however, the spirit cannot penetrate through the form and manifest itself.

When one reflects again on the concept that reincarnation signifies the metamorphosis of the human being, one can now answer the questions: What is the polarity? What is in this case the *Steigerung*? The answers must be: The polarity of the human metamorphosis is past and future. The *Steigerung* is evolution.

This implies that evolution applies only to the human being. Here, the contrast with the actual scientific opinion is very great indeed. We should not forget that the so-called evolutionists did not succeed in putting together an acceptable theory of evolution in a few centuries' time. An acceptable theory would explain how the development of the kingdoms "one out of the other" in a straight line would be without a second thought, comprehensible and creditable. On the contrary, the problems have piled up and grown more acute, however much people originally had hoped for the opposite.

The assertion that the three kingdoms surrounding us have not participated in the evolutionary process is supported by the fact that evolution, as discussed above, can apply only to a self-conscious being. Only a self-conscious being can have talents; only a being that has talents can develop itself; a being that *is* a talent cannot. In this respect, the thought expressed may be especially difficult for animal lovers, who feel that an animal has consciousness also. This is, nevertheless, not self-consciousness!

One may indeed ask what organizes that grandiose animal life? The answer must be that just as the human being has an "I" that is conscious of itself, so too does the entire animal species—but not each separate animal specimen. One could call this "I" a "group-I" or a group soul. One must imagine that this group soul does not live upon the earth, as it does in the case of the human being, but as a spiritual being in the spiritual world. Thus in its life on earth, the animal reveals the activity of the group soul.

To what degree the animal kingdom can be seen in the right light through an entirely different insight into the evolutionary process will be treated further in Chapter VI. Then we shall also speak of the remaining kingdoms of nature.

Chapter 5
The tripartite human being

Three worlds in the human figure

Let us visualize once more what actually happens after death. The trunk detaches itself from the mortal remains; the head remains behind. Our principal activities on earth are connected with the trunk. After death this shape is taken back, enlarged, assimilated, played upon; all this happens in a world we call heaven. At the end of the period between death and the beginning of the next life the trunk is reduced, rolled up, and transformed into a new shape. It receives, so to speak, a stamp from heaven. The new head contains, therefore, the secret of the past and shows the heavenly stamp in the magnificent dome of the skull. Creative beings add a new trunk and the development continues.

Here we are reminded of the build of a newborn child and its development. Is it not striking that the head, which later will seem so small, forms the largest part of the body at birth? "Once the head is through," one hears at a birth. This means that the newborn baby is nearly all head! How different from the adult figure where the head is small in proportion to the rest of the body. While at birth the rest of the body is little more than an appendix to the head, this slowly develops in the true sense of the word to its full size, directed toward the earth. It is worthwhile to become acquainted with the extraordinary laws shown during this further development.

In the first place the baby not only *is* "nearly all head," he *behaves* like a head by letting himself be carried about. As a toddler, he crawls in his playpen. Finally, he stands upright and this introduces a whole new phase. We will restrict ourselves here to the description of the changes in the shape of the body.

Take a look at children less than six years of age. Pathological cases apart, most infants have a harmonious build. They often have strikingly well-shaped legs, in sharp contrast to later years. When one sees a small child in the nude who usually walks beautifully straight, one notices there is no hint of a waistline. The thorax is undeveloped, but the belly is so round and predominant that it often worries inexperienced mothers. All this changes in the seventh year when the arching chest and the waist appear.

At this point there is an interesting change in the way children play. They now begin to play really by themselves. Before the seventh year children play together mainly ring games, in which a child detaches for a moment from the circle, often with a certain shyness, and then returns to the safety of the ring. Building, drawing, painting, playing with dolls, and so on, all so important to small children, are more an occupation than a game.

Games in the true sense begin around the time a child goes to primary school: kites, hoops, tops, knucklebones, marbles, leapfrog, taking aim, throwing—even fighting. This kind of playing has been severely restricted by modern circumstances. Those who remember a time when there were few automobiles will recall with nostalgia a world that still had room for these kinds of games.

With puberty the third element of our body, the limbs, begin to come into their own. Now the child receives his actual arms and legs that, however, he must first learn to master. It seems as though these are offered to him from within the earth, because only now the child develops a full relationship to gravity. This is the time when children don't know what to do with their limbs, the time of slouching. Playing in the old sense of the word comes to an abrupt halt. It is replaced by sports and gymnastics.

This little sketch was necessary to familiarize the reader with the idea that in these three stages of life the human personality and body are closely related to three worlds. Everywhere in one's body, but also within one's entire inner being, one carries heaven and earth. As a being with thoughts, feeling and a will, one lives in one's threefold body: head, chest, and limbs. It is as this unit that one passes time and again during one's life on earth through the phases of evolution.

The union, the joining of those worlds, this linking by the human being of above and below, finds its expression in something, by virtue of which the human shape differs entirely from the animal, any animal, as we shall see. This something can be found in the way the skull rests on the spine, and also in the way the feet carry our body.

The skull rests in almost perfect balance on the spine. The occipital hole in the skull lies a little behind the middle of the base of the skull; in a very small child a little further to the front. However, in animals (including apes) it lies significantly more toward the back.

Now we can see this special relationship of the horizontal base of the skull and the vertical line of the spine is repeated at the other pole of the body. Here the lower leg rests on a dome. In the stages of evolution, the arched foot is a particularly human characteristic. The point is not to ascertain whether an animal exists whose foot is not completely flat, but rather the impression given by the position of the body vis-a-vis the arch of the foot. Here again, the line of the body is perpendicular to the line of an arch. To me these perpendicular connections are very significant. They announce two different worlds. Above is the world of our head, the world of observation and thought, where the human being is consciously open to his surroundings. At the other pole is the world we call earth, the world the human being links up with at birth. In between is the domain where a person is most himself, as when he indicates himself by placing his hand on his chest; for example, when he asks "Do you mean *me*?"

These three worlds live in every person. One could describe them simply as heaven, humanity, and earth. Plates 91 and 92 show how the human form can be absorbed in the dynamics of lines described here.

To understand the relationship of the arched foot with the earth on which it rests, we have to return to Chapter 3 and recall what has been said about the ten little heads on fingers and toes in connection with the build of the long bones and the "change of direction" before and after the carpus and the tarsus.

Let us first look at the foot. What is the consequence of the fact that we have an arched foot? It is that we are not pressed on the earth, that we can have an entirely different relationship with the earth than if we were resting on it with a flat sole.

What does this express? That the sole of the foot is tripartite: a point of contact in which the power of sensation is concentrated, the toes; the heel on which we support ourselves; and, in the middle, that wonderful arch where man is hypersensitive, where he rises, and where he again has something that is strictly his own.

If our foot rested completely on the ground, we would be completely subject to the earth. As things are, the possibility has been created for something that many people suspect, but rarely realize.

Our toes, the heads of our feet, touch the earth with a certain degree of attention. Tribes who walk barefoot show a remarkable sensitivity in scanning the earth with their feet. The middle part of our foot provides yet another kind of observing.

If, apart from considering our feet merely as organs of locomotion, we ask "What do feet observe?", we might think of a scene that often occurs in our lives, in one form or another. A young man said to his mother after dinner,

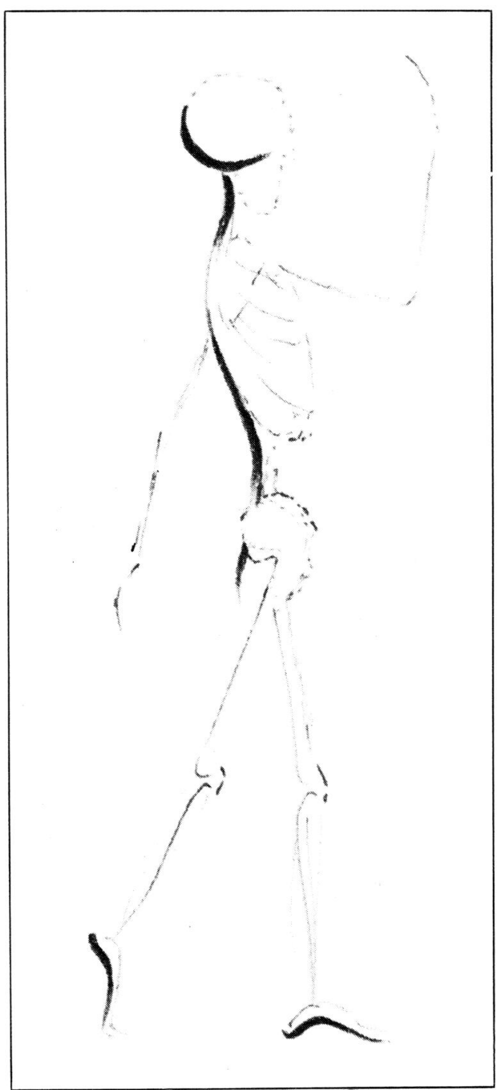

Pl. 91. Drawing illustrating linear dynamics between head, spine, leg, and arch of foot.

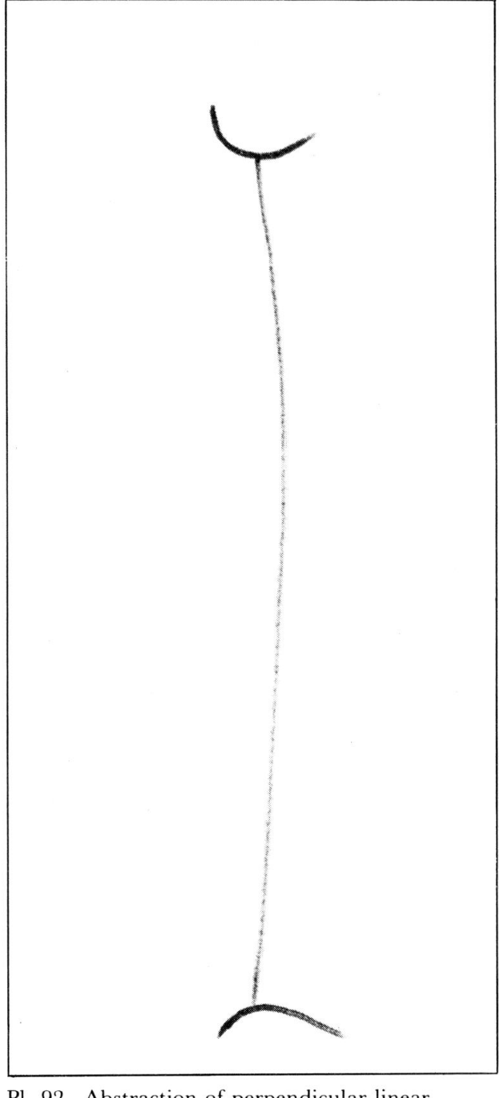

Pl. 92. Abstraction of perpendicular linear relationships in Pl. 91.

"I'm going for a stroll to smoke a cigarette." He quietly walked along the path, turned at a corner and bumped into a young woman. There were profuse apologies on both sides. A conversation followed: "Do you live in this neighborhood?" "How long have you been living here?" The young man liked her, she rather fancied him, they planned to meet again, and, to make a long story short, they had six children.

Because of that one cigarette, remember, or perhaps not? Is there in everybody an unconscious something that is wiser than one's own intelligence? What is the origin of the phrase "to seek one's way, to find one's way?" Perhaps the notion of karma is not a mere fabrication, after all. How often do we hear the phrase "it had to be?" Even the expression "it fell to his lot" may point in that direction. Could there not be wisdom somewhere that far surpasses our own ability in the art of calculation?

Just as we spoke of the feet, we must speak of our hands. Our hands are not flat either; they possess a hollow. This indicates, besides action, observation. We "see" with our hand. Groping implies a kind of seeing—in our pockets to find whether we have the car keys, for instance. But there is something entirely different that corresponds in a striking way to our statements about the feet. We refer to what happens when we shake hands. During the handshake our fingers feel another's skin and the whole hand feels the strength of the handclasp. But there is one place where we do not touch each other—the hollow of our hand. Here, too, we find a structure reminiscent of the middle foot where a person remains himself and where he shows the same sensitivity as in the foot. Here, too, are the thumb, the ball of the little finger, and the heel of the hand with which and on which one can support oneself.

In connection, these ideas help one discover how important it is that we do not cling to each other. When we feel the hand of a small child snuggling completely against ours, we are aware the child's hand does not yet have a hollow. One can clearly feel there is something missing that will come later. Only with children does one play patty cake.

The foot of the small child does not have a hollow, either. Babies are not flat-footed. They have flat feet. The arches in hand and foot appear later, when the child begins to develop consciousness of his surroundings. This shows how long it takes the human being to arrive on earth.

How expressive are handshakes; how different they can be! We have the impression that perhaps unconsciously they play a part in future relationships, and this links up with the impressions of eye and ear. Here lies the secret of every encounter.

What we observe with our everyday senses—our senses of vigil, one might say—has always to do with something that already exists, that has already originated. Thus we can say that in principle our world of observation, our head, always deals with the past. The world of our feet is a different matter. The way that has been described leads into the future.

The question arises: Do not our hands live pre-eminently in the present, do they not make us primarily people of today? Although these distinctions are never quite exact, we can nonetheless see the human being as a link between past and future. The concepts thinking, feeling, and willing can also be spontaneously linked with this image.

In this total image of the human being in his triplicity our hands occupy a very special place. When we said that the path, the life of our feet, always leads into the future, the thought may have arisen that the same goes for our hands, our actions. This is true, but we have also spoken of "seeing" with our fingers.

Where we can speak of seeing with our hands, they belong to the head. In action, they belong to the world of the feet. In an encounter, they belong to our social life. This enables us to see our hands in an entirely new light. In a remarkable way they represent the total human being in past, future, and present; in observation (touch), action, and encounter. The hand's special gift of observation in an encounter, which we can also consider as a kind of seeing, helps us to understand yet another phenomenon in our body.

First, we must note that hand and eye support each other constantly. We describe with our fingers what we have seen; we point to the object we are looking at. The same kind of connection, this time with the foot (even with the entire leg), can be found in a totally different organ of sense: our auditory organ.

As we have seen, the leg is a limb with a clear triplicity. In a surprising way we find the same principle in the chain of auditory ossicles. Here, too, we find a triplicity: hammer, anvil, and stirrup. The hammer is fixed to the eardrum, the stirrup to the internal ear, on the oval window. The whole chain lies in the middle ear, the tympanic cavity. Plates 93 and 94 show these separately and connected together.

If we look at the hammer, we see a relatively big head, a neck, and a pointed small handle that bears a small projection next to the head (pl. 95). Recalling the upper part of the thighbone, we may find a resemblance in that the angle between the neck of the thighbone and the thighbone itself is almost equal to the angle between the neck of the head and the handle of the hammer. Their oscillation is given as being between 125 degrees and 130 degrees.

As one can see, in the hammer everything has gone to the side of the head, in the thighbone toward the limb side. Hammer and thighbone are interesting examples of metamorphosis in the limbs and the head.

When we look at the head and ear of someone with a beard, we are struck by the fact that the ear is situated, so to speak, in an upwards spur of the trunk. The skin beneath the ear and around it is bald. Thus one can get the impression that the ear, apart from being a sense organ, is also a trunk element in the skull, a limb principle in the world of the sense organs (pl. 96).

The reverse must also be considered: Could it be that there is a kind of hearing in the feet, just as we have found a kind of seeing in the hands?

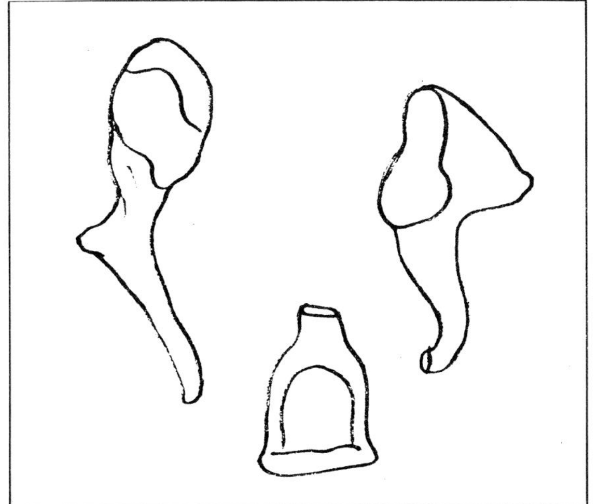

Pl. 93. Hammer, stirrup, and anvil in inner ear (from Cunningham's anatomy textbook *Leerboek der anatomie*).

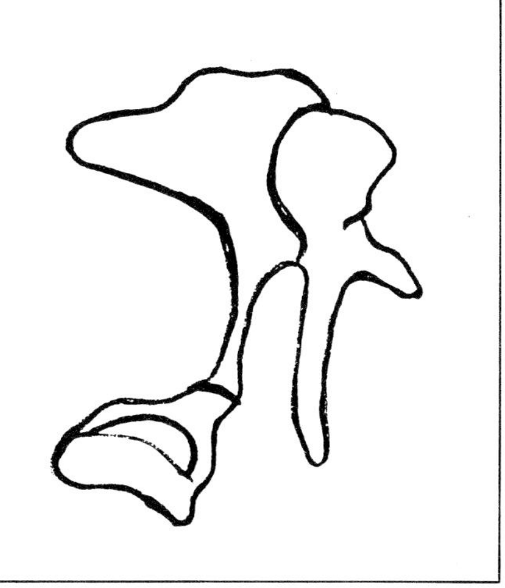

Pl. 94. Hammer, stirrup, and anvil connected as in inner ear (from Cunningham's anatomy textbook *Leerboek der anatomie*).

Could we say that when we find our way through life and listen to our karma, we *listen to* what has been destined for us?

There is a reason for making these suggestions in the form of questions. These suggestions are on a plane to which the human soul must aspire in order to find a new relationship to the problems of life.

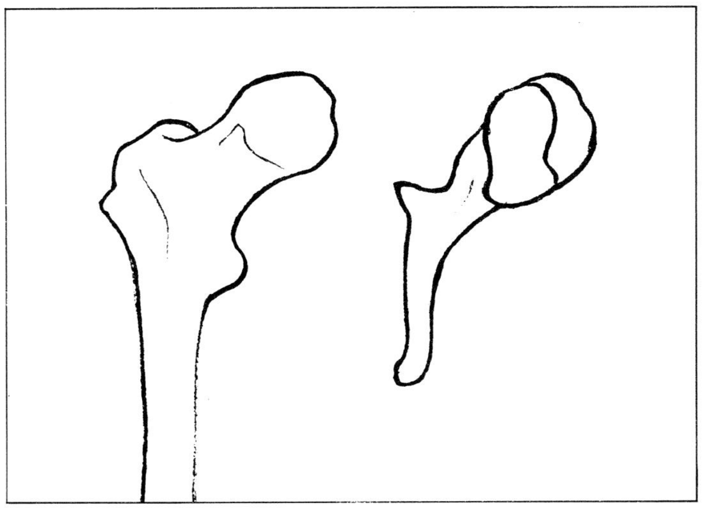

Pl. 95. Upper thighbone and hammer of inner ear.

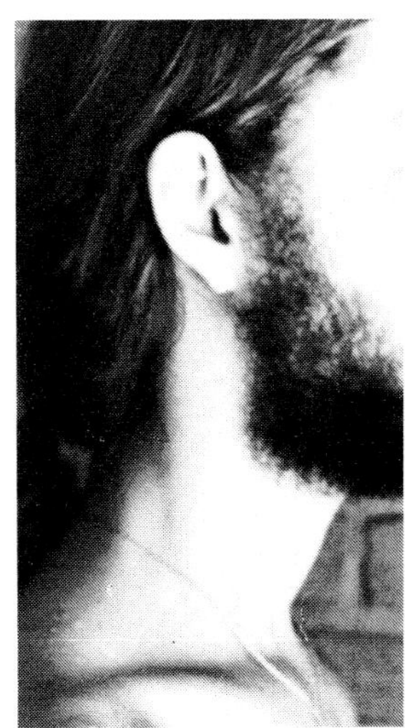

Pl. 96. Placement of the ear.

The threefold composition of the skull

Since we are dealing not only with the central part of the skull, but also with the facial cranium and with the vault, we must discuss the shape of the entire skull. The vault consists of a number of flat bones, which do not possess a shape in the same sense as the shoulder blade, the temporal bone, and so on. One could say that the face has disappeared (see the section "Thoracic vertebra and shoulder blade" in Chapter 3).

In order to continue our train of thought about the metamorphosing of the trunk into the skull, we have to ask ourselves which bones are re-encountered in the vault, which bones have been transformed into the vault. We have found in the central skull (temporal bones and sphenoid bone and two jaws) the girdles and limbs (metamorphosis of vertebrae and ribs) of the axial skeleton.

The bones of the vault—especially the frontal bone, parietal bones, occipital bone—are situated where we can imagine the ribs in the bent-forward trunk. They must be pictured in their entirety with the ribs attached, but completely flattened and smoothed out.

In returning to the facial cranium and observing the many holes and inlets the profile shows in connection with the sense organs, we may ask ourselves what is actually the characteristic of these sense organs. Generally speaking, it is best to attack a problem where it shows itself most clearly. The obvious choice for a discussion on the characteristics of the sense organs, therefore, is the eye.

An eye is often compared to a camera. What is the basis of this great similarity? Someone with knowledge of photography can answer the following questions: Why does one need a lens? Without a lens the light does not form concentrated pictures. Why is the interior of the camera dark? If it wasn't dark, light would be reflected and create a halo. Why is there a shutter? Without a shutter the light would act too long on the photosensitive substance. Why is there a diaphragm? Without a diaphragm, the light cannot produce sharply defined pictures.

All these answers have the same subject: light.

It is not difficult to realize that the camera and the eye have been manufactured and produced by man and by the body, respectively. Both eye and camera have been constructed according to the laws of light. Light dictates the method. Goethe was right when he said that the eye has been created by light for light. This can help us gain insight into the question of the origin of the sense organs in general. There is no clearer example of similarity than the eye and the camera. The eye can show us a special way to attack the riddle of the sense organs.

Using our findings about the eye as a point of departure, we can say that the sense organs are created, brought forth, by something that itself can then be observed by them. These resemble apparatuses because in part they follow the laws of the outside world and in part they are the body's answer to the outside world. Rudolf Steiner spoke of "inlets" of the outside world. This can be said of *every* sense organ; *in principle* it is always true.

Because most sense organs are situated in the head, the human being is more or less directed forward, and we can distinguish three areas in the skull. First, there is a central part that I would like to call *limbs-human being-skull* in connection with the metamorphosis we have described. It consists mainly of temporal bones, the sphenoid bone in between, and the jaws and cheekbones.

Second, there is *sense organ-earth-skull*, the area that has been transformed by the sense organs as they have just been characterized, allowing what is outside to enter in the shape of earthly impressions. The same is true for the middle skull, of course, with the auditory organ.

Third, there is the *vault-heaven-skull*, which encloses our brain and blends into the two preceding areas without any sharp demarcation. The sense organs link up with the world; this part may therefore be called the earth side of the skull. The vault is the cosmic side. In the middle is the essentially human part that links the other two principles together. It is in the nature of the discussion to point to the similarity between the threefold structure of the body and the newly discovered triplicity in the form of the skull.

The relationship of the earthly and cosmic to the human side of the skull changes in the course of life. This can be explained in the following manner. In plate 97 the earthly and cosmic principles are each indicated by a dotted line.

In a very young child, a baby, the part face-sense organ is hardly developed vis-à-vis "heaven." In a young person this part gradually increases. It

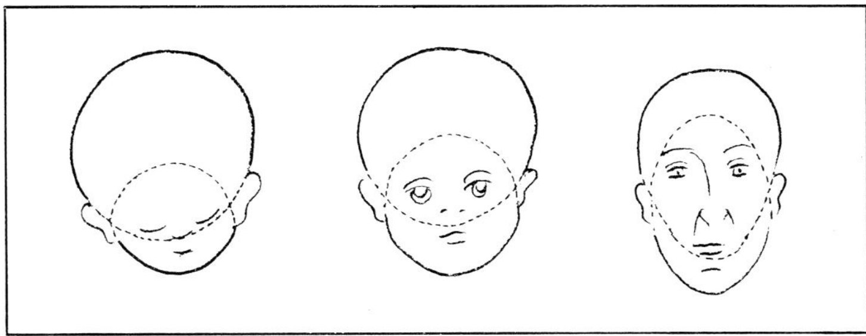

Pl. 97. Heads of baby, youth, and adult in terms of "heavenly" and "earthly" parts of skull.

Pl. 98. Animal head with earthly part of skull predominating.

is only in older people that we see how the human being orientates himself over time increasingly toward the earth.

By placing these examples next to the fourth, in which the earthly element is exaggerated, (Pl. 98), we see the striking way the animal becomes apparent. In the animal the earthly element predominates from the start of its existence. Plate 99 shows a human skull and the skull of a monkey, as a clear example of what has been said here.

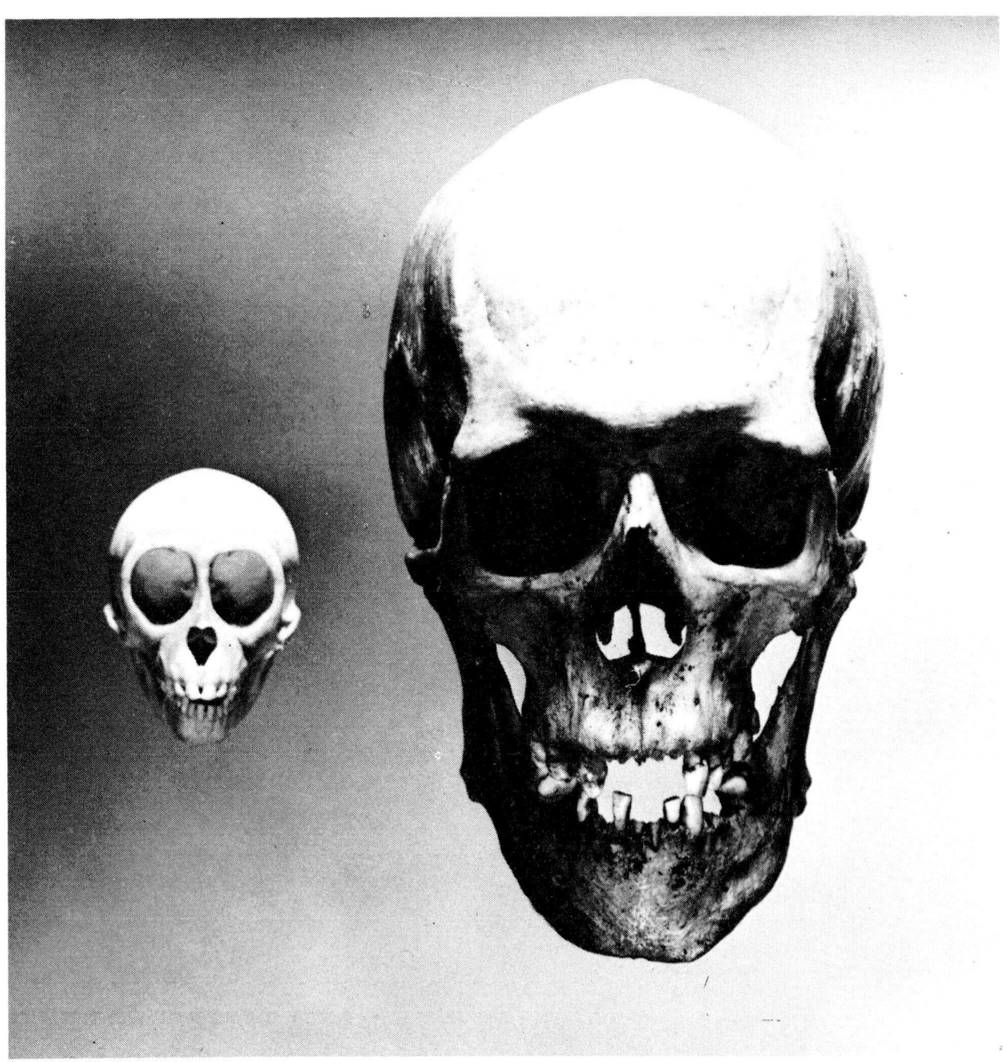

Pl. 99. Skulls of monkey and human being.

Our cosmic form

Recalling the point of departure of our research, we are reminded that first we found our theme in the metamorphosis of the human skeleton—namely, the composition of vertebra with ribs.

Before that, we had as theme of the plant metamorphosis, the principle of leaf with stalk.

The word theme is a term used in music and other forms of art. In the theme of a piece of music we hear the concentrated form of the work, which will be elaborated later. The content of such a theme strikes us by its richness.

Where does the theme spring from? Here we often use the word inspiration, which really only indicates the way of receiving. The origin itself lies in the expression "a divine theme." In our matter-of-fact times this is considered merely an emotional term. The Greeks thought differently; they considered all forms of art to be revelations of divine beings they called the Muses. Thus music, for the Greeks and for many of us the most direct revelation of that divine world, is honorably named.

What revelation is there in the scheme of our skeleton? To find this out we should study the vertebra-with-ribs shape from above. We then see two enclosed cavities: at the back the spinal canal and in the front the arch of the ribs, including the sternum (pl. 100).

If we follow in our imagination the line of the first cavity or arch in an upward direction, it becomes bigger and bigger until it reaches the cerebral cavity with which it merges. In the other direction it narrows down and reaches a dead end in the coccyx.

The front arch does the opposite. Upwards, it ends in the smallest arch of the first rib. It widens in the other direction, until the arch opens up in front at the floating ribs.

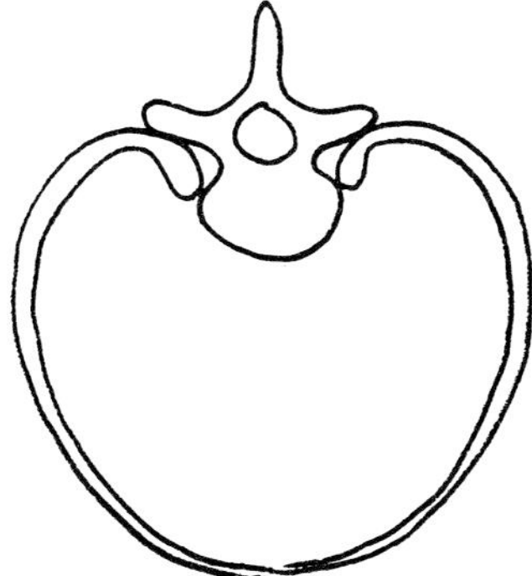

Pl. 100. Vertebra-with-ribs shape seen from above.

It will not be easy for everybody to accept the idea that the opening arch then directs itself toward the earth and continues in our legs. By allowing oneself to be saturated with the ascending and descending metamorphosis of the shapes of these lines, one may arrive at a new approach to this idea. One begins to understand that it is closely linked with the triplicity of our total form, and valid only for the human being.

The animal skeleton shows magnificent metamorphoses, especially in the spine with the ribs. If, however, one sees how the tail and the head lie in an almost direct line with the horizontal spine, one feels the essential difference from the threefold human form and its upright position.

This thought has been rendered by line drawings in plate 101. The arms are also indicated as a third element, representing a world of its own that has its own place between the two contrasts. If one sees what was said about the hands as relevant to this, it will become clear why this addition is justified and even necessary.

It is even possible to go a step further and to study the lines in connection with the shape of a lemniscate. Mathematically it is possible to construct this in many ways; under certain circumstances it can appear as two circles. What we have found in plate 101 as ascending and descending shapes can also be seen as the metamorphosis of a lemniscate.

It is well known that the planets move in special orbits around the sun. Seen from the earth, these orbits show a more or less complicated system of nodes, loops, etc. These lines can, mathematically speaking, be reduced to one prototype; here, too, we meet the lemniscate. We may say that the planets move in metamorphoses of lemniscates.

One might object, of course, that these are illusory movements and the planetary orbits are seen in that shape only from the earth. This is true, but it does not alter the fact that this so-called illusory movement is a reality to us and that the impression made on us by such a movement must be interpreted according to that illusion. The lemniscate shape that can be found in our body is also valid only for our body on earth. Just as the illusory lemniscate presents a different aspect in the world of the planets, so will the human lemniscates reveal a different aspect before birth or after death in a world with different dimensions.

Some years ago a friend of mine sent me a very charming illustration of the human form as lemniscate in the shape of a vignette. It is the form of a Buddha composed of three metamorphoses of a lemniscate (pl. 102). The principal element of the theme of the human form, the vertebra in the shape of a lemniscate (pl. 103), is offered here as a finishing touch.

Pl. 101. Drawing showing contrasting gestures of head and limbs, with arms between.

Pl. 102. Vignette of Buddha composed of metamorphosed lemniscates.

Pl. 103. Drawing of vertebra in shape of a lemniscate.

So we arrive at the conclusion that the human figure expresses in the shapes of his skeleton that he is a cosmic being. He lives on earth as a temporary guest; on earth he creates his biography. In his figure he carries a cosmography. This means that we must seek the true origin of the human figure where we discovered the source of inspiration in music.

Would it not be possible to view the skeleton as a symphony? True, the skeleton is a spatial object we can see, but it is also possible to hear it as a stillness from the harmony of spheres, as Kepler mentioned in his astronomical works.

We mentioned numerical ratios in the discussion about the structure of arms and legs as spreading lines. The intervals in music are also numerical ratios.

Rudolf Steiner has created a new art of movement, eurythmy, that makes music visible in movements. The musical intervals are indicated by upper arm, wrist and hand in such a way that the series of numbers we encountered there (1-2-3-4-5) express at the same time a series of intervals (first, second, third, fourth, fifth).

These are very vague indications, but to my mind they are permissible in the context of a description that, starting from the exact, tries to identify itself in the world of the "in-act," the world from which these shapes once originated and still originate.

Chapter 6
Metamorphosis and evolution

The various metamorphoses we have studied in plants, animals, and humans will now be juxtaposed so we may discover new connections.

What was the theme of the plant metamorphosis? A form-theme that can be characterized as a leaf. The typical way the total shape of the plant comes into being is that form after form emerges, one after and above the other.

The animal metamorphosis is entirely different, as we have seen. Each phase is absorbed by the following one. It was necessary to find another theme for this metamorphosis as form would not do, and we found desire. Desire acts in the urge to become what finally emerges; all that precedes is absorbed therein. This is the big difference from the plant.

Looking at the various individual phases through which the animal metamorphosis passes, we again find the image of the metamorphosis we have come to know in the plants. We find this in the segments of the lower animals (worms), in the articulations of the arthropods, in the spine of the higher animals. This appears to be a plantlike metamorphosis, but it does *not* come about by joining part to part. Quite the contrary, these series of shapes emerge in their entirety from the undifferentiated. We can see it in a simple way in the caterpillar. Its complete ring structure is present from the beginning and only grows in size. The metamorphosis of the human skeleton is in this sense, too, principally a metamorphosis of form, and therefore a plant metamorphosis. Here also the segmentation of the vertebrae slowly emerges as a *totality* in the embryonic development.

Just as we encounter images of plant metamorphosis in the animal, so we can say that an animal metamorphosis can be found in the human being. The development from ovum into a fully developed human being can be compared fully with that of the animal. The human form is something all people have in common and is described in books of anatomy and physiology.

The actual human metamorphosis that we have described as reincarnation entails yet another transformation. To be able to accept this, it was necessary to find in each human being something that distinguishes itself in principle from everything that comprises the animal. We refer to Chapter 2.

We touched on a principle that is as much the theme of reincarnation as desire is the theme of the animal and form is the theme of the plant. The animal evidently has the plant metamorphosis behind it, the human being both plant and animal metamorphosis. As an evolving being, his own metamorphosis lies between past and future.

We shall now go one step further and develop the idea of metamorphosis in regard to the animal kingdom as a whole. The animal kingdom is divided into a number of groups, with which most of us are familiar: the series of lower and higher animals, the invertebrate and the vertebrate, the cold and warm blooded, etc. The contrast between lower and higher animals that slowly appeared on earth one after the other can be found in classifications such as earlier and later; simple and complicated; those strongly linked with the environment and those with a far greater independence from it (for example, mammals). Polarity and *Steigerung* (intensification) are much in evidence here.

Confronted with the question whether this is a metamorphosis, one has to accept the idea that each higher animal species is, therefore, the metamorphosis of a lower one and has developed from it. This is precisely what the adherents of the traditional evolutionary theories have always maintained. We must declare emphatically that they have not succeeded in proving this. No idea in this direction has anything to do with real experiences. Quite the contrary. The conviction that the animal kingdom is an *image* of a metamorphosis, but is not one *in actual fact* is gaining more and more ground.

It is exactly the same with the plant kingdom. Here too are earlier and later plants, primitive and complicated ones, etc.; and here too we find a sequence of species forming a clear series. Still, since it can never be proved that, for example, toadstools have turned into higher plants (although there is a clear case of polarity and *Steigerung*), one is confronted with the same enigma: We have the image of a metamorphosis, while the reality escapes our experience.

To complete our view we add the mineral kingdom. We must juxtapose the four kingdoms of nature—mineral, plant, animal, and human—knowing that in principle they have appeared or at least started to appear one after the other.

The recognition of polarity and *Steigerung* in the image of these four kingdoms belongs to the most fascinating chapters of our study. In the case of the mineral kingdom we must first establish the fact that it possesses nothing of what we attribute as life, soul, or spirit to plant, animal, or human being. The plant lives; the animal, in addition to living, has feelings and desires; the human being carries something that makes him capable of self knowledge.

Again one cannot help but speak of polarity and *Steigerung.* But can it be said that a mineral has changed itself into a plant, a plant into an animal, and, consequently, an animal into a human being? True, one still tries very hard to learn to see, think, and experience it this way, especially since the image of evolution runs so clearly through the whole chain. We are indeed faced here with a great dilemma: the certainty of an evolution and the equal certainty that it is impossible to maintain the conventional idea of everything arising from what has preceded it.

To understand still better the phenomenon of metamorphosis, we will briefly mention the characteristics of each kingdom.

A region of earth void of human, animal, and plant is called a desert. Like the French word *desert*, it means forsaken. Remember the expression God-forsaken; forsaken or abandoned means that once the situation was different, once there was a creative principle. By applying this to the mineral realm we render the origin of this mineral earth at least as plausibly as the primal explosion (big bang) theory that is the modern point of departure. We must choose between "the Word," Kepler's music of the spheres, or the primal explosion.

What characterizes mineral substances? How does one distinguish them? By their properties. How do we know these properties? Most are known through research. Physical and chemical properties are discovered mainly in laboratories. This means at first sight they are hidden. Only in special circumstances created by humanity (living beings!), such as dissolving and heating, do they begin to show their many properties.

This enigmatic realm becomes even more striking if we ask ourselves what the inside of a piece of matter looks like, since all we can see is the surface. Even molecules, atoms, and electrons—each of the suggested parts of matter—show us only their outside, their surface. Returning to our point of departure, we discover the puzzling fact that matter is only surface. Precisely because we are dealing with an object of a certain dimension it is so difficult to realize that when all is said and done, matter is nothing but surface. This means it has no inner self, and justifies the expression "matter reflects properties." That is all a surface can do—reflect.

The mineral realm is the hidden reflection of properties of creative beings, beings that have abandoned it (see Mees, *Living Metals*, London: Regency Press, 1974).

Turning to the realm of the plants, one notices how much it differs from the mineral world. Here there is no question of hidden properties, everything is out in the open without anybody's intervention.

What a wealth of images are presented to us in the course of the year! The seasons unfold before our eyes primarily through the plant world. The plant consists essentially of stem and leaves, lines and planes. Apparent exceptions do not alter this principle.

The plant is quite receptive to its surroundings. Warmth, light, air, water, and minerals—everything the plant needs for its growth—is taken directly from the environment and absorbed. It is right to say that plants link heaven and earth. But one can also say that they *live* between heaven and earth. Such life, such unbridled vitality, is at work in this world. Plant life is growth, plant growth is life, one might say. Thus we could describe the plant as *an open image of life between heaven and earth.*

The animal is best characterized by a new element, the desire that brings with it an inner world. Here for the first time we can speak of a life of soul, but limited by sensory impressions. Related to this is the fact that the animal form develops in an entirely different way, especially in higher vertebrate animals.

There is no greater contrast imaginable than that between an animal and a plant. I like to say: Animals have an interior, plants have an exterior. Thus we know where we are.

It has already been mentioned that every animal species has its own desire. This is what we recognize in animals, expressed and visible to the finest detail. Let us take a look at animals and see what is so fascinating. First of all the colors and the covering, such as scales, shells, armor, feathers, fur, and so on. Fascinating, too, are those amazing appendages: antlers, horns, fangs, tusks, trunks, claws, hooves, wings, beaks, and fins.

There is no better way of describing an animal than by saying that it is completely *expressed.* The animal's environment is also decided by its appearance. A fish not only can swim, it *must* swim. An animal, therefore, lives only in that environment in which it can satisfy its desire. Thus we came to the conclusion in Chapter 2 that *the animal is a closed unity of desire, shape and environment.*

And now the human being. Those who consider him merely an animal must eliminate ideas like responsibility, morality, and freedom. Many modern anthropologists, being consistent, do just that.

For others this viewpoint denies our real humanity, which is still at the beginning of its development. For them *the human being is the revelation of the divine evolution on earth.* How must we see this?

The disclosures of Rudolf Steiner in anthroposophy concerning evolution may help give us an answer. He does not give a theory or an idea, but speaks of personal experiences that can be shared by anyone prepared to develop organs of observation in a field that has fallen outside our routine, everyday consciousness. I say "has fallen," because in the past people had access to the realm that we call the spiritual world. That power of observation has left us; what matters is the future development of new organs (see Chapter 4).

What is the fruit of research done by someone who can read in the "book of memory of the development of the earth" (the Akasha Chronicle)? Namely, that it is true that man has never been a mineral, plant, or animal, but that he has passed through those phases in three consecutive planetary situations. This was concurrent with the condensation of matter-warmth-air-water into earth. During each phase a realm was created to free evolving humanity from too strong an inclination toward condensation (mineral), too great a vitality (plant), and too strong a pressure of desire (animal). This is the origin of the realms of surrounding nature. It need hardly be said that they have undergone enormous changes over millions of years due to their own development. In the image they present today we can see their origin, however, and what we have sought: the evolution of the human being.*

All this is linked together by the evolution of humanity. At several points (phases) the evolving being was obliged to reject the other realms so that he would not be brought to a standstill.

This makes the series of realms so striking. We can go even further in our characterization by saying that the creative-spiritual can be recognized consecutively *out* of the mineral, *at* the plant, *in* the animal, *through* the human being (see Mees, *Living Metals*).

We can recapitulate this in a simple outline. We have mentioned the various stages of human evolution. The different kingdoms found on earth form an uninterrupted chain and represent the mineral, plant, animal, and human phases.

The specific characteristic of each realm, and therefore of each phase, is the possession of respectively, body, life, soul, and spirit. These are given by the world of creative beings, the world of hierarchies. All this has taken place in condensing matter, respectively warmth-air-water-earth.

It is very important to view the whole in such a way that the difference between the first three phases vis-a-vis the fourth, the human phase, is mani-

*The paragraph I quoted from Professor Bolk in Chapter 2 shows he, too, believed human evolution was decisive for the creation of the animal world.

fest. We should speak of three preparations. It is only in the last phase that fulfillment, in the truest sense of the word, takes place.

In the evolution of the human being all that has been said so far about the mineral, plant, and animal kingdoms finds its context. We have said that plant and animal realms, in their many sorts and shapes, form a chain that gives a clear *image* of evolution.

Every plant and animal species has led an independent existence from its inception. In the course of ages these have, naturally, undergone great changes, but they have never attained a higher form.

The dilemma we were faced with now starts to come into perspective. In thinking more deeply about the notion that these kingdoms are segments of human evolution, many things become clear. One can imagine that the creative events took place in the least condensed regions of the earth. Evolution came about the way a work of art does. Once it has been shaped, once the shape has slipped into its outline, further metamorphosis is out of the question. At the same time this can make clear what significance the birth of the kingdoms of nature has had for human evolution. It was essential that the human form develop further and not prematurely assume a definite contour.

Only the prehistoric shapes are known to us as the origin of the shapes of plants and animals. They are the only link in the chain of evolution lying

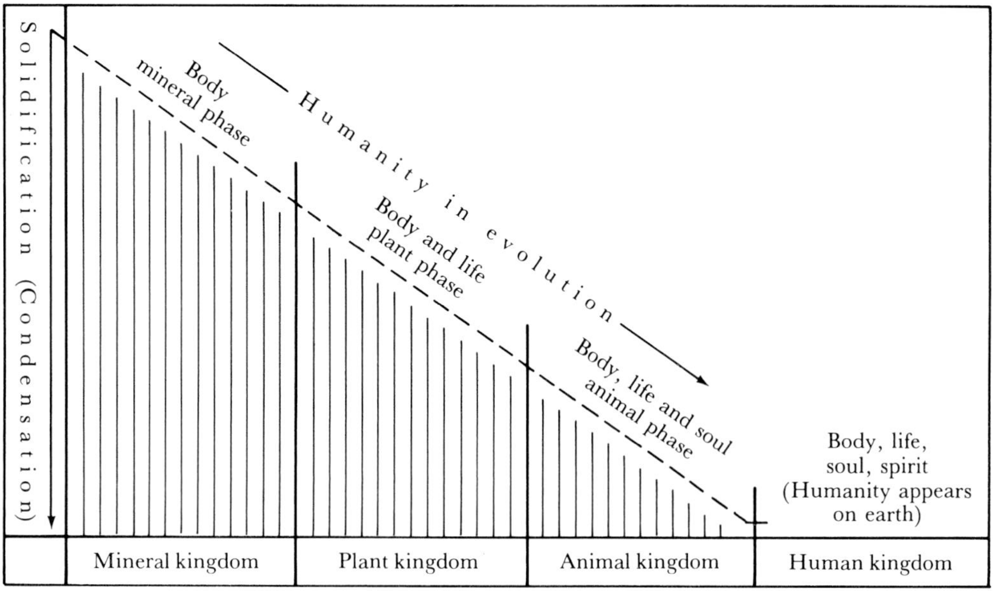

*It should be emphasized that the events described in Rudolf Steiner's *Occult Science* are far more intricate and complicated. The chart is just a simple diagram to indicate the essential.

within our experience. When those prehistoric shapes changed into our modern plants and animals, the possibility for further changes in shape became very restricted. Only variations were and are possible. The origin of the prehistoric shapes belongs to a certain phase of the evolution of the human form.

Thus it becomes understandable that we find the plant metamorphosis back in the animal bodies, as we have seen, but now as one that from the beginning presents itself as a whole. We have said that the animal has passed beyond the plant metamorphosis, but this is not altogether correct. It is not the animal form itself that has passed through a phase; it is the evolving human being passing through the animal and plant metamorphosis.

The animal carries the plant metamorphosis in itself; the human being, before going through the animal phase, went through the plant phase. The human being has gone through the animal phase and we find in him the animal metamorphosis.

In this way we make a triple acquaintance with metamorphosis. First, the metamorphosis in the plants and animals around us, after which we discussed human metamorphosis. Second, the *image* of a metamorphosis that we find in plants and animals, each as a total realm. Third, the image of a metamorphosis presented by all four kingdoms together. Now it is possible to find a satisfactory connection between the various experiences.

In Chapter 1 it was said that the human being recognizes himself as a creative being. With him the creative force appears on earth. If one would consider those preparatory stages as parts of a divine evolution, one might conclude that this evolution flows into the human evolution and is a continuation of it. Hence the phrase: Humanity represents the revelation of the divine evolution on earth.

The question arises: What does it all mean? Is it, however, possible for us to answer such a question? To ask for the meaning is meaningless. The question should be: What is my task in my surroundings? This question can be asked concerning every element in nature. Perhaps it would be better to word it like this: What role does a particular thing play in its surroundings?

One can only speak of a task when a being possesses self-consciousness, when that being is capable of setting himself a task. A century ago all thought of suitability (teleology) had already been excluded from botany and zoology. This was correct because only a self-conscious being can really have a task, an aim.

It is clear that what we have just stated can only apply to something in development. Therefore one cannot say that the human being should behave in such and such a way, that he is this or that. Man *is* not, he *becomes.* There is no end in sight. Evolution can never have a limit. With the human there appears for the first time a being carrying something within him with which he faces his surroundings. The creatures of nature, animals included, do not face their surroundings—they experience them. People can think about their surroundings; these present an "image," leaving them, in principle, free.

We must not forget for one moment that here we are dealing with a very subtle principle that lives potentially in every person, but as yet is only at the beginning of its development. Consequently, we can follow human evolution through history. Until the time of Greece human evolution was in preparation. There were large groups that had to be led and guided by special persons in order to reach the stage in which they could think about their surroundings. Only then the slow and gradual development of the special quality that is so typically human could begin. The leaders were not ordinary people. They were well ahead of their time and possessed a trained consciousness of the spiritual world. We do not have space to elaborate on this subject, but it helps to remember that such initiated persons were called demigods for good reason.

It was not until ancient Greece that science, whose development we still experience, came into being. It paved the way for the materialistic outlook on life of the West, which rejected all previous values as so many illusions. It would be well to realize that this attitude holds a certain temptation! The materialistic attitude holds a certain attraction because, up to a point, one may speak of a desire for security.

What is the task that has been reserved for the human being? It must be a task to which he can say yes, something that can come into being only after a free relationship with the surroundings is accomplished. It is then that we begin to discover a design in the history of humanity.

Minerals, plants, and animals are accomplished facts, results of creative occurrences. They represent the work of creative beings who, unlike humanity do not appear on earth. Therefore they are excluded from evolution. As we have already seen, they merely offer an image of evolution. Evolution takes place on earth, in the human being.

Only the human being possesses a consciousness that can deny the divine. Therein lies his strength and also a danger. In each human being lives a seed placed in him that enables him to take a new direction of development, a direction in which his role is not one of denying, but of recognizing, the divine in his environment. Then he can become conscious of his unique position.

Humanity alone is capable of thinking about what has been created; indeed, thinking represents a new element in evolution. It provides the human being with an independent spiritual life and enables him to develop moral attributes. Morality can be considered only when it is linked to responsibility. The availability of freedom is indissolubly tied to this. In this way alone do moral attributes obtain their significance. It is even understandable that these attributes can originally be denied. We can reject the notion that the human is a programmed computer, a being without responsibility who has to adapt himself to imposed precepts (behaviorism); but it should not surprise us that such ideas do exist and even find a large audience. Without this risk the acquisition of a new understanding would depend not on strength or a victory but on a conclusion. Thus it would, in fact, lose its value.

If a well-defined point of view is described here, it is based on manifold experiences in which anthroposophy has played a great role. A point of view can also produce a perspective. Broadly speaking, we are facing a panorama that over time can lead to better discernment and new insight. We might describe it this way: Between birth and death, the human being lives in this world with a consciousness *facing* creation. But the discernment he develops regarding his evolution is, at the same time, something that is experienced in the creative world by higher beings. One might say that in the life between death and birth the human being lives in a world that asks him a question.

The originally hidden answer lies in life on earth. This life can become the possibility for us to be involved again in the evolution of the surrounding kingdoms (to which we owe our self-conscious earthly life) and to take these kingdoms with us in our future development. This is my viewpoint; I can say "yes" to it.

Such thoughts often encounter a strong opposition from religion. The origin of our train of thought and the development of the idea and the phenomena we have discussed are rarely entered into. Religion simply declares that all this is based on overestimation of oneself. What is the human being but a speck of dust in the great universe? What is he in comparison with God who rules all? Is there not a hidden self assertion, and possibly vanity, behind this outlook?

I would like to reply to this with a statement capable of countering the opposition and showing some understanding for the attitude of contemporary youth towards religion. In the course of centuries a conviction has taken hold in the Christian church which will very reluctantly be relinquished. This is the idea that God is perfect. To me, this means *in essence* that God is dead. A perfect being is, in a sense, dead.

One could also try the thought that God is the most imperfect being imaginable. This thought understandably provokes resistance because we are inclined to equate imperfect with bad and perfect with good. This is not at all our intention. We are suggesting the possibility of viewing the divine as open to infinite development. And it offers the possibility to think that this development continues in human evolution. Some will say, "What arrogance!" I say rather, "What a responsibility!" The justification of this responsibility is really what anthroposophy is all about.

If it is true that we live in apocalyptic times, then the appearance of the writings of Antoine de St. Exupery can be seen in a particular light. For those who know his work, the following may perhaps form the conclusion of the train of thought described here.

It has been repeatedly said how the actual core of man is, as yet, quite delicate. We have spoken about a principle in its beginning; we might say a small principle—"a little prince."

In one of the first chapters of *The Little Prince*, St. Exupery says: "I want my book to be taken seriously." It contains a warning to guard ourselves against the loss of our true humanity in the whirlpool of technical, rational life. The last sentence that the fox (earth consciousness) teaches the Little Prince is: "You are responsible for your flower." Does not the word "responsible" indicate the highest treasure of humanity?

Is it not striking that behaviorists emphatically deny responsibility?

Is it not through responsibility that man actually begins to be human?

Conclusion

Perhaps it would be well to state that what we have found to be the metamorphosis of the shape of the axial skeleton in the shapes of the skull must in no way be considered proof of Rudolf Steiner's declarations. Such things can never be proved. The most one can do is to find coherent images that produce such thoughts. As with all proof, one can consider the things Rudolf Steiner has communicated to us from within his own insight as proven only if one has come to know them through personal experience. In this case that applies only to those who are prepared to school their consciousness in the ways discussed.

Repeatedly seeing and touching the various parts of the skeleton provides us with a growing certainty. This is not only due to familiarity with the shapes, but is accompanied by a growing enthusiasm. Slowly things become clear; we begin to see something—better still, to read something.

Regarding each of these comparisons of form, one might wonder if they are not just a matter of coincidence. This changes as soon as we connect all these coincidences, and we begin to see that together they form an organic whole. It is the beginning of a realization.

This is closely linked with the way one is induced to study these forms. There is always an artistic side; it is the natural consequence of studying forms. The artistic element in our observation can never be sufficiently appreciated. It opens our eyes to worlds other than those of sensorial exactness, to which our modern scientific thinking is inclined to restrict itself. That is why it is an indispensable element in the development leading to a transcendental observation.

Thus one can indicate the line in the development of the human consciousness as follows: What once was reality for humanity became faith, what can again be reality must become foreboding. Faith always points to the past, foreboding to the future. They meet each other in the present, as understanding.

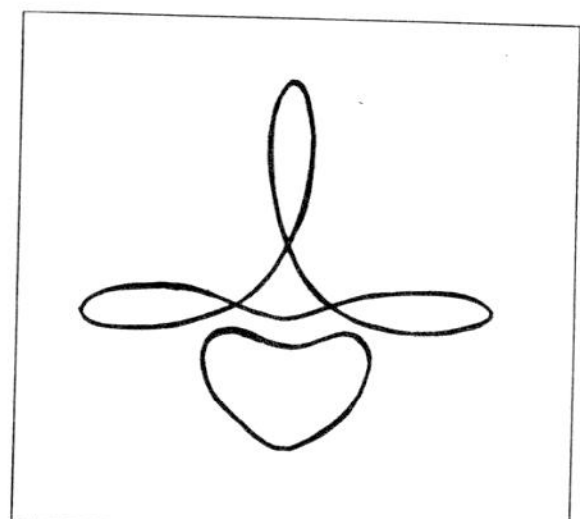

A Note from SteinerBooks

SteinerBooks is a 501(c)(3) not-for-profit organization, incorporated in New York State since 1928 to promote the progress and welfare of humanity and to increase public awareness of Rudolf Steiner (1861–1925), the Austrian-born polymath writer, lecturer, spiritual scientist, philosopher, cosmologist, educator, psychologist, alchemist, ecologist, Christian mystic, comparative religionist, and evolutionary theorist, who was the creator of Anthroposophy ("human wisdom") as a path uniting the spiritual in the human being with the spiritual in the universe; and to this end publish and distribute books for adults and children, utilize the electronic media, hold conferences, and engage in similar activities making available his works and exploring themes arising from, and related to, them and the movement that he founded.

- We commission translations of books by Rudolf Steiner unpublished in English, as well as new translations for updated editions.
- Our aim is to make works on Anthroposophy available to all by publishing and distributing both introductory and advanced works on spiritual research.
- New books are publish for both print and digital editions to reach the widest possible readership.
- Recent technology also makes it efficient for us to make our previously out-of-print works available for the next generation.

SteinerBooks depends on our readers' financial support, which is greatly needed, appreciated, and tax-deductible. Please consider a donation by check or other means to SteinerBooks, 610 Main St., Great Barrington, MA 01230. We also accept donations via PayPal on our website. For more information about supporting our work, send email to friends@steinerbooks.org or call 413-528-8233.